Mechanical Science IV

ERRATA

Page 7(b)ii : $SF_{xx} = wl - wx$ not $wl - \frac{wx}{2}$

Page 33 and 34 : reverse captions for Figs 2.9 and 2.10

Page 36 Fig.2.11(b) : Stress distribution at section XX'

Page 66 Table 3.1(a)i : maximum deflection is $Wl^3/3EI$

Page 89 Fig.4.22(a) : plane AD not shown

Page 92: Transpose lines 14-20 with lines 1-13.

Page 139 Fig.6.18(a) : 143 not 145 rev/min

Page 168 Fig. 8.5 : note that axis ZZ' (not shown) is perpendicular to the plane of the paper.

Mechanical Science IV

M. A. Rix

HODDER AND STOUGHTON
LONDON SYDNEY AUCKLAND TORONTO

M. A. Rix ACP CEng MIMechE was formerly Senior
Lecturer in Mechanical Engineering at Richmond
upon Thames College.

Cover photograph by courtesy of the Hawker Siddeley Group Ltd.

British Library Cataloguing in Publication Data

Rix, M.A.
   Mechanical science IV.
   1. Mechanics, Applied
   I. Title
   621.1   TA350

   ISBN 0 340 36785 7

First printed 1985

Copyright © 1985 M. A. Rix

All rights reserved. No part of this publication may be reproduced or transmitted
in any form or by any means, electronic or mechanical, including photocopy,
recording, or any information storage and retrieval system, without permission in writing from the publisher.

Typeset in 11/12 pt Univers (monophoto) by Macmillan India Ltd, Bangalore.

Printed in Great Britain for
Hodder and Stoughton Educational,
a division of Hodder and Stoughton Ltd.,
Mill Road, Dunton Green, Sevenoaks, Kent TN13 2YD.
by J. W. Arrowsmith Ltd, Bristol

Contents

Preface	vi
1 Loading, Shear Force and Bending Moment Diagrams *(Standard unit 4.1)*	1
2 Stress due to Bending *(Standard units 2.1, 2.2, 3.1, 3.2)*	26
3 Deflection of Beams *(standard units 5.1 to 5.7)*	47
4 Stress and Strain *(Standard units 1.1 to 1.16)*	70
5 Belt Drives *(Standard units 6.1 to 6.5)*	106
6 Mechanisms and Velocity Diagrams *(Standard units 7.1 to 7.4)*	123
7 Flywheels and Turning Moment Diagrams *(Standard units 8.1 to 8.6)*	146
8 Vibrations *(Standard units 9.1 to 9.7)*	163
9 Balancing *(Standard units 10.1 to 10.4)*	188
Answers to Exercises	200
Index	209

Preface

This book is written to cover the objectives of the Business and Technician Education Council's standard unit U82/041, but it could also be of use to students on many courses covering similar topics. The relevant sections of the standard unit covered by each chapter are indicated in the contents pages. Although Chapter 1 contains a large amount of material normally covered by a student at Level III it is included for the benefit of students who enter the Higher BTEC courses from other courses of study.

A series of self-assessment questions are included at the end of each chapter. These refer to the theory and they will have greater value if, initially, they are attempted without reference back to the text. Reasoned answers, where appropriate, are given at the end of the book and large number of unworked examples are included at the end of each chapter. These are designed to be of a standard required for phase or end tests.

In connection with Chapter 6, where graphical solutions to examples are required, such solutions are printed to a much smaller scale than would be required to ensure reasonable accuracy. In these cases guidance is given with regard to the minimum actual scale which is recommended for accurate solutions.

The abbreviations and symbols for units, and the units themselves, are as far as possible those recommended by the BSI and Business and Technician Education Council.

1 Loading, Shear Force and Bending Moment Diagrams

1.1 Introduction

The term 'beam' is normally associated with structures, although machine components, pipes carrying fluids etc., can often act as beams and they must be designed accordingly. The theory concerned with this design, *bending theory*, depends for its application, upon a knowledge of the maximum bending moment as well as the beam dimensions.

1.2 The diagrams

Loading diagram. This is a sketch or a scaled drawing of the beam indicating the nature and the positions of the supports and the loads. It is usually drawn prior to the other two diagrams.

Shear force diagram. This is a sketch or a scaled drawing showing the variation in the shear force from section to section of the beam due to the particular loading.

Bending moment diagram. Since the bending moment varies according to the positions and types of load, their combined effect is portrayed by a scaled drawing or sketch called a bending moment diagram.

1.3 Bending moments, shear force definitions and sign convention

Bending moment. This is the total moment, at any section of a beam, due to all the forces *either* to the right of the section *or* to the left of the section.

Shear force. This is the algebraic sum of all the forces *either* to the right of the section *or* to the left of the section.

Sign convention. A bending moment may have a clockwise or an anticlockwise sense and a shear force may be in one of two directions, so it is necessary to adopt a consistent method of distinguishing between them. The convention which is to be used is illustrated in Fig. 1.1. It should be noted that other conventions, if consistently applied, are equally valid.

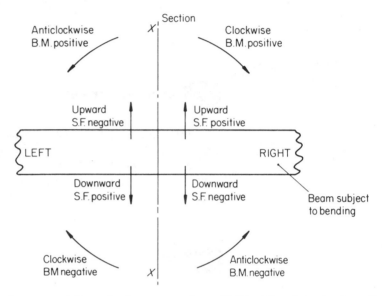

Fig. 1.1 Bending Moment (B.M.) and Shear Force (S.F.) sign convention

At any section of a beam, such as XX' in Fig. 1.1, the bending moment and the shearing force can be calculated by considering all the forces to the *right* of the section and ignoring those to the *left* . Then:

> a *clockwise* bending moment is *positive*
>
> an *upward* shear force is *positive*.

If the forces to the left of the section are used, and those to the right are ignored, then:

> an *anticlockwise* bending moment is *positive*
>
> a *downward* shear force is *positive*.

1.4 Diagram construction; simple cases

In the examples which follow, the various diagrams are sketched only approximately to scale as any significant values are calculated for accuracy. Loads in newtons are preferred to masses in kilograms since it is the gravity force on the mass which gives rise to the bending moment and shear force.

Example 1.1
A simply supported beam 0.8 m long carries a concentrated central load of 3 kN. Draw the load diagram and sketch the bending moment and shear force diagrams. Determine:

(a) the bending moment and shear force at a section 0.2 m from the right-hand support;
(b) the value and position of the maximum bending moment.

Load Diagram, Fig. 1.2 (a). The load is central so the support reactions are each equal to half the load.

$$\therefore R_1 = R_2 = 1.5 \text{ kN}.$$

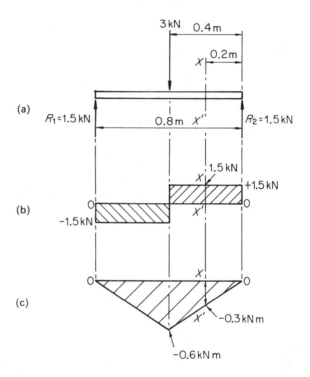

Fig. 1.2 (a) Load diagram
(b) Shear force diagram
(c) Bending moment diagram

Shear force diagram, Fig. 1.2 (b). Working from the right-hand end (RHE), the shear force (SF) at the RHE changes from 0 to +1.5 kN due to the support reaction.

SF at section XX' = +1.5 kN due to the forces to the right of XX'.

Considering the forces to the left of section XX',

$$SF = +3 \text{ kN} - 1.5 \text{ kN} = +1.5 \text{ kN}.$$

The SF, therefore remains constant from the RHE since there are no other forces between section XX' and the RHE.

SF at the centre = +1.5 kN, changing abruptly to −1.5 kN due to the 3 kN concentrated load at this point.

The shear force will then remain constant at −1.5 kN up to the left-hand support when it changes abruptly to zero due to the support reaction.

Bending moment diagram, Fig. 1.2 (c). The BM (bending moment) at the RHE = 0 since there are no forces to the right of this. If the forces to the left are considered then:

$$\text{BM at RHE} = +(3 \times 0.4) - (1.5 \times 0.8) = 0 \text{ as expected.}$$

Considering the forces to the right of XX',

$$\text{BM at section } XX' = -1.5 \times 0.2 = -0.3 \text{ kN m.}$$

Considering forces to the left of XX',

$$\text{BM at section } XX' = (3 \times 0.2) - (1.5 \times 0.6) = -0.3 \text{ kN m.}$$

Considering either the forces to the right or to the left of the centre,

$$\text{BM at the centre} = -(1.5 \times 0.4) = -0.6 \text{ kN m.}$$

The bending moment is obviously proportional to the distance from either end and it is a maximum under the centre load.

From this simple example it should be apparent that:
(i) between concentrated loads or forces the bending moment increases uniformly and the bending moment diagram is a sloping straight line;
(ii) between concentrated loads or forces, the shearing force remains constant but changes abruptly at those loads or forces;
(iii) the bending moment has a maximum value where the shear force changes from + to − or passes through zero.

The next example illustrates the differences in both the BM and the SF diagrams when the loading is evenly distributed.

Example 1.2
A beam is simply supported at its ends over a span of 0.8 m and carries a distributed load of 3 kN/m. Draw the load diagram and sketch the bending moment and shear force diagrams. Determine and indicate the value of the bending moment and shear force at:
  (a) a section XX' 0.2 m from the right-hand end;
  (b) at the centre of the beam.

Load diagram, Fig 1.3 (a). Since the load is evenly distributed it will be shared equally between the supports.

$$\text{Total load} = 0.8 \times 3 = 2.4 \text{ kN}$$

$$\therefore R_1 = R_2 = \frac{2.4}{2} = 1.2 \text{ kN.}$$

Shear force diagram, Fig 1.3 (b). Working from the RHE:

SF at RHE = 0 changing abruptly to +
  1.2 kN due to the reaction R_2

Considering the load to the right of XX'

$$\text{SF at } XX' = +1.2 - (3 \times 0.2) = 0.6 \text{ kN.}$$

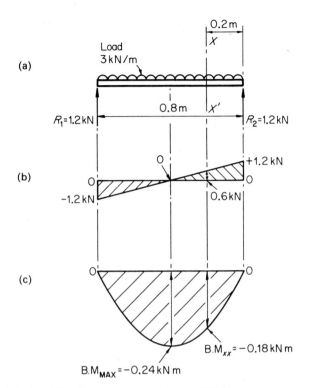

Fig. 1.3 (a) Load diagram
(b) Shear force diagram
(c) Bending moment diagram

Considering the forces to the left of XX',

$$\text{SF at } XX' = -1.2 + (3 \times 0.6) = 0.6 \text{ kN.}$$
$$\text{SF at centre} = +1.2 - (3 \times 0.4) \text{ or } -1.2 + (3 \times 0.4)$$
$$= 0$$

SF at LHE $= -1.2$ kN, changing abruptly to zero.

Bending moment diagram, Fig. 1.3 (c). Working from the RHE:

BM at RHE $= 0$ since there are no forces to the right of R_2.

A similar result is obtained by considering the forces to the left of R_2.

$$\therefore \quad \text{BM at RHE} = -(1.2 \times 0.8) + \left(\frac{3 \times 0.8^2}{2}\right)$$
$$= -0.96 + 0.96 = 0$$

Considering the forces to the right of XX',

$$\text{BM at } XX' = -(1.2 \times 0.2) + \left(\frac{3 \times 0.2^2}{2}\right)$$
$$= -0.24 + 0.06 = -0.18 \text{ kN m.}$$

Considering the forces to the left of XX',

$$\text{BM at } XX' = -(1.2 \times 0.6) + \left(\frac{3 \times 0.6^2}{2}\right)$$
$$= -0.72 + 0.54 = -0.18 \text{ kN m.}$$

Considering the forces to the right of centre,

$$\text{BM at centre} = -(1.2 \times 0.4) + \left(\frac{3 \times 0.4^2}{2}\right)$$
$$= -0.48 + 0.24,$$
$$= -0.24 \text{ kN m.}$$

A similar result is obtained by considering the forces to the left of centre. The bending moment no longer varies linearly with the distance from either end but in proportion to the distance squared. The bending moment diagram is thus sketched as a curve between the points.

From this second example it is possible to conclude the following points.

(i) The shear force for an evenly distributed load varies directly with distance from one end and the diagram is thus a sloping straight line. With end supports this line passes through zero at the centre.

(ii) The bending moment for an evenly distributed load varies with the square of the distance from one end and the bending moment diagram is a curved line.

(iii) The bending moment is a maximum at the centre of the simply supported beam where, more significantly, the shear force passes through zero i.e. it changes sign.

1.5 Standard cases

Examples 1.1 and 1.2 used specific loads and beam dimensions. However, as these types of loading are commonly encountered it would be far more useful to develop the solutions in general terms. They then become known as *standard cases*. Five such standard cases are summarised in Table 1.1.

It is worthwhile noting from Table 1.1 of standard cases that both the shear force and bending moment can be expressed in terms of the load (w or W), the length of the beam (l) and the distance of a beam section from one end (x). The ability to write the expressions in this way, particularly the bending moment, becomes very important later when calculating the deflections caused by different loads.

Table 1.1.

Bending moment and shear force diagrams: standard cases.

(a) Simply supported beam; centre load

(i) *Load diagram*

$$R_1 = R_2 = \frac{W}{2}$$

(ii) *Shear force diagram*

$$SF_{MAX} = \mp \frac{W}{2}$$

$$SF_{XX'} = \mp \frac{W}{2}$$

(iii) *Bending moment diagram*

$$BM_{MAX} \text{ (at the centre)} = -\frac{W}{2} \times \frac{l}{2} = -\frac{Wl}{4}$$

$$BM_{XX'} = -\frac{Wx}{2}$$

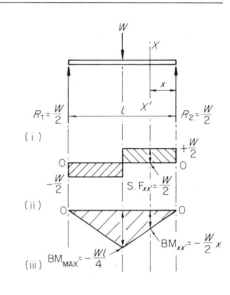

N.B. In this case when $x > \frac{l}{2}$ the BM is calculated from the opposite end of the beam.

(b) Simply supported beam; distributed load

(i) *Load diagram*

$$R_1 = R_2 = \frac{wl}{2}$$

(ii) *Shear force diagram*

$$SF_{XX} = +wl - \frac{wx}{2}$$

$$SF_{MAX} = \mp \frac{wl}{2}$$

(iii) *Bending moment diagram*

$$BM_{XX} = -\frac{wlx}{2} + \frac{wx^2}{2}$$

$$BM_{MAX} = -\frac{wl^2}{4} + \frac{w}{2}\left(\frac{l}{2}\right)^2$$

$$= -\frac{wl^2}{8} \text{ when } x = l/2$$

$$-\frac{Wl}{8} \text{ if } wl = W.$$

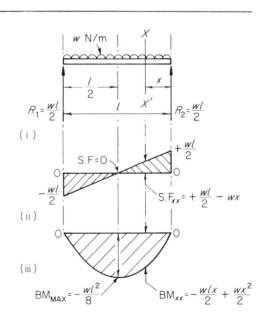

(c) Simply supported beam; concentrated offset load

(i) *Load diagram*
Taking moments about R_1:
$$Wa = R_2(a+b)$$
$$R_2 = W\frac{a}{(a+b)}$$

Taking moments about R_2,
$$Wb = R_1(a+b)$$
$$R_1 = W\frac{b}{(a+b)}$$

(ii) *Shear force diagram*
Working from the RHE.
For values of $x < b$:
$$SF_{xx} = +R_2 \text{ or } +W\frac{a}{(a+b)}$$
For values of $x > b$:
$$SF_{xx} = -R_1 \text{ or } -W\frac{b}{(a+b)}$$

(iii) *Bending moment diagram*
Working from the RHE.
For values of $x < b$:
$$BM_{xx} = -R_2 x$$
For values of $x > b$:
$$BM_{xx} = -R_2 x + W(x-b)$$
BM_{MAX} occurs at the load
$$= -R_2 b \text{ or } -R_1 a.$$

(d) Cantilever; single concentrated end-load

(i) *Load diagram*

(ii) *Shear force diagram*
$$SF_{xx} = -W \text{ (Constant)}$$

(iii) *Bending moment diagram*
$$BM_{xx} = +Wx$$
$$BM_{MAX} = +Wl$$
at the support when $x = l$.

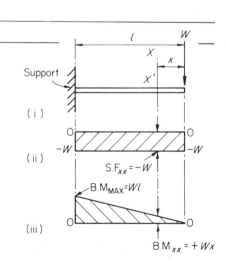

(e) **Cantilever with distributed load**
  (i) *Load diagram*
  (ii) *Shear force diagram*

  $SF_{xx} = -wx$
  $SF_{MAX} = -wl$
     (at the support)

  (iii) *Bending moment diagram*

  $BM_{xx} = +\dfrac{wx^2}{2}$

  $BM_{MAX} = \dfrac{wl^2}{2}$ or $\dfrac{Wl}{2}$

  where $W = wl$, and occurs at the support.

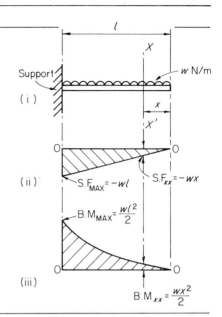

1.6 General cases; mixed loads

In practice, beam loadings will not always consist of a single concentrated or distributed load. The weight of the beam itself may constitute the distributed load, to which may be added concentrated loads at various points. To determine the bending moment and the shear force distributions in such cases a more general treatment is necessary.

Example 1.3 Two concentrated loads
A horizontal shaft is simply supported (in spherical bearings) over a span of 0.5 m, and carries two gear wheels of weight 200 N and 400 N at distances of 0.1 m and 0.3 m from the right-hand end respectively. Ignoring the weight of the shaft, determine:
  (a) the shear force at each load;
  (b) the bending moment at each load;
  (c) the maximum bending moment, and where it occurs.
Sketch, approximately, the *load, shear force* and *bending moment* diagrams.

Load diagram, Fig. 1.4(a). Taking moments about R_2,

$$0.5R_1 = 0.1 \times 200 + 0.3 \times 400$$
$$R_1 = \frac{20 + 120}{0.5} = 280 \text{ N}$$
$$R_2 = (400 + 200) - 280 = 320 \text{ N}.$$

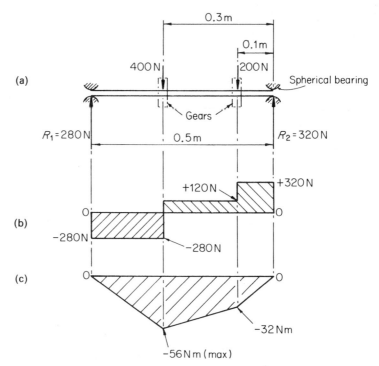

Fig. 1.4 (a) Load diagram
(b) Shear force diagram
(c) Bending moment diagram

Shear force diagram, Fig. 1.4(b). At the RHE, where the reaction R_2 occurs, the shear force changes abruptly from 0 to +320 N. It is then constant up to the 200 N downward force where it will change abruptly to +120 N, i.e. +320 − 200. The shear force is once again constant up to the 400 N load, where it changes abruptly to −280 N i.e. +120 − 400. It then remains constant as far as the left-hand support where a further abrupt change to 0 N occurs i.e. −280 + 280.

Note that the shear force passes through zero at the section where the 400 N load is applied.

Bending moment diagram, Fig. 1.4(c). Working from the RHE, at R_2 the BM = 0. At the 200 N load section the BM due to the forces to the right of this section

$$= -R_2 \times 0.1$$
$$= -320 \times 0.1$$
$$= -32 \text{ N m}.$$

Since all the loads are concentrated, the change in bending moment between the two points on the diagram is uniform. At the 400 N load section, considering the forces to the right of this section,

$$BM = -R_2 \times 0.3 + 200 (0.3 - 0.1)$$
$$= -320 \times 0.3 + 200 \times 0.2$$
$$= -56 \text{ N m}.$$

Once again, because the loads are concentrated, the change from −32 N m to −56 N m is uniform as shown on the diagram. At the left-hand support, considering all the forces to the right,

$$\text{BM} = -R_2 \times 0.5 + 200(0.5 - 0.1) + 400(0.5 - 0.3)$$
$$= -320 \times 0.5 + 200 \times 0.4 + 400 \times 0.2$$
$$= 0.$$

From -56 N m to the left-hand support the bending moment changes uniformly to zero.

The maximum bending moment is thus -56 N m and it occurs at a section under the 400 N load. This is also the section at which the shear force *passes through zero*. As suggested by the standard cases there is, apparently, a connection between the section where the maximum bending moment occurs and the section at which the shear force passes through zero (i.e. changes sign).

The next example illustrates the effect of adding to the concentrated loads an evenly distributed load representing the weight of the shaft.

Example 1.4 Mixed loads

If the shaft in Example 1.3 was of uniform cross-section and had a weight of 100 N determine once again:

   (a) the shear force at each load;
   (b) the bending moment at each load;
   (c) the maximum bending moment and where it occurs.

Sketch the approximate load, shear force and bending moment diagrams.

Load diagram, Fig. 1.5(a). The shaft weight is evenly distributed and it is equivalent to $\dfrac{100}{0.5}$ or 200 N/m. Taking moments about R_2:

$$0.5 R_1 = 0.1 \times 200 + 0.3 \times 400 + \frac{(200 \times 0.5^2)}{2}$$

$$R_1 = \frac{20 + 120 + 25}{0.5} = 330 \text{ N}.$$

$$R_2 = (400 + 200 + 100) - 330 = 370 \text{ N}.$$

This, as is to be expected, shows that half the shaft weight is added to each reaction as calculated in Example 1.3.

Shear force diagram, Fig. 1.5(b). At R_2 the shear force changes abruptly from 0 to $+370$ N. Between R_2 and the 200 N load it will decrease uniformly by 200 N/m, due to the distributed load (i.e. the weight of the shaft). At the 200 N load, therefore,

$$\text{SF} = +370 - (0.1 \times 200) = 350 \text{ N}.$$

At this same section the shear force then changes abruptly to 350 -200 or 150 N due to the effect of the 200 N load. Between the 200 N and 400 N loads the same uniform decrease due to the distributed load will occur so that at the 400 N load,

$$\text{SF} = 150 - (0.2 \times 200) = 110 \text{ N}.$$

The abrupt change due to the 400 N load itself causes the shear force to then become 110 − 400 N or − 290 N at the same section. At the left-hand support,

$$SF = -290 + (0.2 \times 200)$$
$$= -330 \text{ N}.$$

This then changes abruptly to zero due to the effect of the reaction R_1.

Bending moment diagram, Fig. 1.5(c). Working from the RHE, at R_2 the BM = 0 (No forces to the right of R_2). At the 200 N load section, due to the forces to the right of the section,

$$BM = -R_2 \times 0.1 + \left(200 \times 0.1 \times \frac{0.1}{2}\right)$$
$$= -370 \times 0.1 + \left(200 \times \frac{0.1^2}{2}\right)$$
$$= -36 \text{ N m}.$$

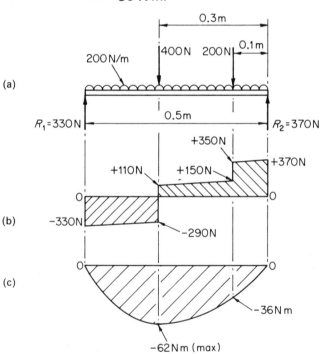

Fig. 1.5 (a) Load diagram
(b) Shear force diagram
(c) Bending moment diagram

At the 400 N load section, again considering the forces to the right of the section,

$$BM = -370 \times 0.3 + 200(0.3 - 0.1) + 200 \times \frac{0.3^2}{2}$$
$$= -111 + 40 + 9 = -62 \text{ N m}.$$

At the left-hand support, the BM should be zero since there are no forces to its left. However, as a check consider the forces to the right:

$$BM = -370 \times 0.5 + 400(0.5 - 0.3) + 200(0.5 - 0.1)$$
$$+ 200 \times \frac{0.5^2}{2}$$
$$= -185 + 80 + 80 + 25 = 0.$$

In sketching the diagram between the points calculated the joining lines are curved due to the distributed load effect. A more exact curve could be obtained by calculating values of bending moment for some intermediate sections. In the same way, by taking sections just to the right or left of the 400 N load, it can be confirmed that the bending moment at the 400 N load section is the maximum. Again the maximum bending moment occurs where the shear force passes through zero and is -62 N m.

In order to facilitate later work on the calculation of the deflection, the next example will be dealt with in a slightly different manner.

Example 1.5 Use of formulae for BM and SF
A cantilever 1.5 N long has a weight of 3 kN and carries concentrated loads of 4 kN at the free end, and 5 kN at a distance 0.5 m from the free end. Determine:
(a) the shear force at each load and the support;
(b) the bending moment at each load;
(c) the maximum bending moment, and where it occurs.
Sketch the *loading*, *shear force* and *bending moment* diagrams.

Loading diagram Fig 1.6(a). The distributed load due to the weight of the beam
$$w = \frac{3}{1.5} = 2 \text{ kN/m}$$

Working from the free end, which is essential for a cantilever:
(i) For sections distance x from the free end $x \leqslant 0.5$ m. The SF and the BM are due to the distributed load and the 4 kN load only, thus:

$$SF_{xx} = -4 - 2x \qquad (1.1)$$
$$BM_{xx} = +4x + \frac{2x^2}{2} \qquad (1.2)$$

(ii) For sections distance x from the free end the value of x is between 0.5 m and 1.5 m.

$$SF_{xx} = -2x - 4 - 5 \qquad (1.3)$$
$$BM_{xx} = +4x + \frac{2x^2}{2} + 5(x - 0.5) \qquad (1.4)$$

Thus by using the Equations 1.1 and 1.3 the shear force at any section can be calculated, and from Equations 1.2 and 1.4 the BM is calculated.

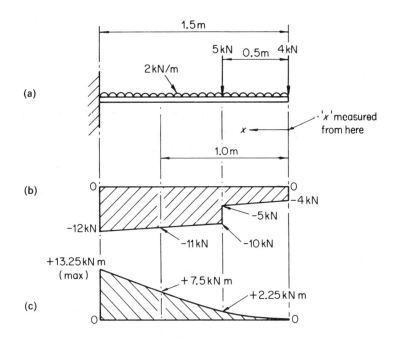

Fig. 1.6 (a) Load diagram
(b) Shear force diagram
(c) Bending moment diagram

Shear force diagram Fig 1.6(b). The SF at the free end = 0 so, using Equation 1.1, when $x = 0.5$

$$SF = -4 - (2 \times 0.5) = -5 \text{ kN,}$$

and from Equation 1.3,

$$SF = -4 - (2 \times 0.5) - 5 = -10 \text{ kN.}$$

Since the 5 kN load is applied at this section, the shear force will change abruptly between these two values. Using Equation 1.3:

when $x = 1.0$, $SF = -4 - (2 \times 1) - 5 = -11$ kN;
when $x = 1.5$, $SF = -4 - (2 \times 1.5) - 5 = -12$ kN.

Bending moment diagram Fig 1.6(c). The BM at the free end = 0, so using Equation 1.2, when $x = 0.5$:

$$BM = +4 \times 0.5 + \frac{2 \times 0.5^2}{2} = +2.25 \text{ kNm}$$

The same result would be obtained by using Equation 1.4 since the additional term $+5(x - 0.5) = 0$ when $x = 0.5$.

$$\text{when } x = 1.0 \; BM = +4 \times 1.0 + \frac{2 \times 1^2}{2} + 5(1 - 0.5)$$

$$= +7.5 \text{ kN m}$$

$$\text{when } x = 1.5 \; BM = +4 \times 1.5 + \frac{2 \times 1.5^2}{2} + 5(1.5 - 0.5)$$

$$= +13.25 \text{ kN m (maximum value).}$$

Further consideration of Equation 1.4 for the bending moment will show that it is identical with 1.2, except for the addition of the $5(x-0.5)$ term. Consequently the single equation 1.4 could be used for calculating the bending moment providing the term $5(x-0.5)$ is only used when the value of the bracket $(x-0.5)$ is positive, i.e. $x > 0.5$. To remember this it is usual for [ ] brackets to be used in such cases. Thus Equation 1.4 is written:

$$BM = +4x + x^2 + 5[x-0.5].$$

Using this technique, therefore, a single equation for the bending moment at any section of a loaded beam can be written. This will prove particularly useful when dealing with beam deflections.

One further example should consolidate the ideas developed above.

Example 1.6 Mixed loading with overhung beam.
A beam 2 m long is simply supported at the left-hand end and at a point 1.5 m from the left-hand end. Its weight is equivalent to a distributed load of 2 kN/m. In addition, it carries concentrated loads of 4 kN at the right-hand end, and 6 kN at a distance of 1 m from this end.
  (a) Calculate the reactions and sketch the loading diagram.
  (b) Determine the SF at the RHE and at distances 0.5, 1.0 and 2.0 m from the RHE; sketch the shear force diagram.
  (c) Derive an equation to calculate the bending moment at any section XX' which is distance x from the RHE.
  (d) Using the equation derived in (c) calculate the BM when $x = 0$, 0.5, 1.0 and 2.0 m; sketch the BM diagram.
  (e) Determine the value and position of the maximum B.M.

(a) *Loading diagram*, Fig. 1.7(a). Calculate the reactions by taking moments about R_1

$$1.5 R_2 = (6 \times 1) + (2 \times 2 \times 1) + (4 \times 2)$$
$$R_2 = \frac{6+4+8}{1.5} = 12 \text{ kN}$$
$$\therefore R_1 = 6 + 4 + (2 \times 2) - 12 = 2 \text{ kN}.$$

(b) *Shear force diagram*, Fig 1.7(b). The SF at the RHE = 0, changing abruptly to -4 kN due to the 4 kN load.
  The SF 0.5 m from the RHE = $-4 - (2 \times 0.5) = -5$ kN which then changes abruptly to $-5 + 12 = +7$ kN due to the support reaction R_2.
  The SF 1.0 m from the RHE = $-4 - (2 \times 1) + 12 = 0$ kN which then changes abruptly to $0 - 6 = -6$ kN due to the 6 kN load.
  The SF at the LHE (i.e. 2 m from the RHE) = $-4 - (2 \times 2) + 12 - 6 = -2$ kN which then changes abruptly to 0 due to the reaction R_1.

16 Mechanical Science IV

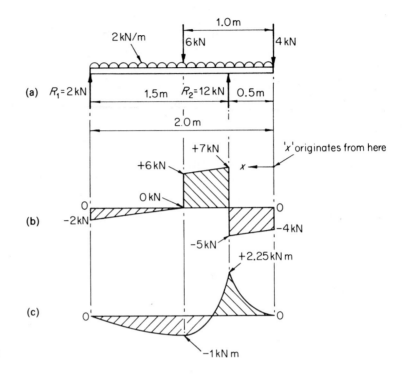

Fig. 1.7 (a) Load diagram
(b) Shear force diagram
(c) Bending moment diagram

(c) *Bending moment diagram*, Fig. 1.7(c). The BM at any section, distance x from the RHE:

when $x \leqslant 0.5$, $\quad$ BM $= 4x + \dfrac{2x^2}{2}$ kN m;

when $x > 0.5 \leqslant 1$, $\quad$ BM $= 4x + \dfrac{2x^2}{2} - 12[x - 0.5]$ kN m;

when $x > 1.0$, $\quad$ BM $= 4x + \dfrac{2x^2}{2} - 12[x - 0.5]$
$$+ 6[x - 1] \text{ kN m.}$$

Using the square bracket notation [ ], the last expression is the equation required for the answer to (c).

(d) When $x = 0.5$, BM $= 4 \times 0.5 + \dfrac{2 \times 0.5^2}{2} = 2.25$ kN m

When $x = 1.0$, BM $= 4 \times 1 + \dfrac{2 \times 1^2}{2} - 12[1 - 0.5]$
$$= -1 \text{ kN m}$$

When $x = 2.0$, BM $= 4 \times 2 + \dfrac{2 \times 2^2}{2}$
$$- 12[2 - 0.5] + 6[2 - 1]$$
$$= 0 \text{ kN m.}$$

Between these calculated values the bending moment diagram is curved due to the effect of the distributed load.
(e) Since the shear force passes through zero at 0.5 m and 1.0 m from the RHE there are two points of maximum bending moment. By examination, the 2.25 kN m is numerically greater and will cause the greater bending stress.

1.7 Point of contraflexure

Example 1.6 suggests that for an overhung beam there are two maxima for the bending moment, one positive and one negative. Between these values, therefore, the bending moment must pass through zero. This point on the beam is known as a point of *contraflexure*. Since the bending effect changes direction so will the slope of the beam, thus at the point of contraflexure the slope will be zero. The slope of a beam is dealt with more fully in Chapter 3.

1.8 Relationship between loading, shear force and bending moment

From the observations made in the previous sections, and the fact that both the shear force and the bending moment occur as a result of the loading, it is to be expected that there is a useful mathematical relationship between them. Establishing this relationship will provide an alternative method of determining bending moments which can be used in cases where the normal methods, previously described, become rather involved.

Fig. 1.8 shows part of a loaded beam, where δx is the length of a small element, $ABCD$, which is some distance from a datum, O. w is the average rate of loading on the element which can be considered as uniform if, as is assumed, δx is very small. On the face AB of the element, F = shear force, M = bending moment. On the face CD of the element, $F + \delta F$ = shear force and $M + \delta M$ = bending moment. The increases at CD are caused by the additional loading between A and C and the increased distance from O.

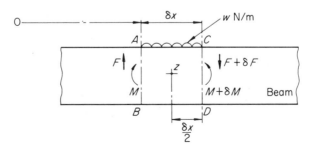

Fig. 1.8 Part of a loaded beam

Since the element is in equilibrium:

(i) the total vertical forces acting on it are zero

$$\therefore F = (F + \delta F) + w\delta x$$
$$\delta F = -w\delta x$$
$$\frac{\delta F}{\delta x} = -w \qquad (1.5)$$
$$\therefore F = -\int w\,\delta x. \qquad (1.6)$$

thus from Equation 1.6, by integrating the expression for the loading, an expression for the shear force can be obtained;

(ii) the total moment on the element is zero.

$$\therefore M + F\delta x/2 + (F + \delta F)\,\delta x/2 = M + \delta M$$

(taking moments about mid-point Z)

$$F\delta x/2 + F\delta x/2 + \frac{\delta F\,\delta x}{2} = \delta M$$

(the small product $\delta F\,\delta x/2$ can be ignored)

$$\therefore F\,\delta x = \delta M \qquad (1.7)$$
$$\text{and} \quad \int F\,\delta x = M \qquad (1.8)$$

Equation 1.8 indicates that by integrating the expression for the shear force an expression for the bending moment can be obtained.

Example 1.7

For a simply supported beam, with an evenly distributed load of w N/m and a length l, derive an expression for the loading, and from it obtain expressions for the shear force and bending moment at any section.

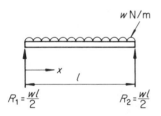

Fig. 1.9 A simply supported beam with an evenly distributed load

From Fig 1.9 w = loading

$$\therefore \text{SF} = \int w\,\delta x = wx + A \qquad (1.9)$$
$$\text{BM} = \int F\,\delta x = \frac{wx^2}{2} + Ax + B \qquad (1.10)$$

A and B are constants of integration which can be evaluated by inserting known conditions *i.e.*

When $x = 0$, BM $= 0$ $\quad\therefore\quad$ from $\dfrac{wx^2}{2} + Ax + B = 0$

$$0 + 0 + B = 0$$
so that $B = 0$.

When $x = l$, BM $= 0$ $\quad\therefore\quad \dfrac{wl^2}{2} + Al = 0$ so that $A = \dfrac{-wl^2}{2l}$

$$= -\frac{wl}{2}$$

By reference to Fig. 1.9 it is obvious that $A = R_1$ or R_2.

Thus from Equation 1.9,

$$\text{SF} = wx - \frac{wl}{2} \qquad (1.11)$$

and from Equation 1.10

$$\text{BM} = \frac{wx^2}{2} - \frac{wlx}{2}. \qquad (1.12)$$

Compare 1.11 and 1.12 with the standard case in Table 1.1(b).

This example demonstrates that the constant A represents the support reaction. From the standard case it is also known that the maximum bending moment occurs when $x = l/2$.

If Equation 1.12 for the bending moment is differentiated and equated to zero the point(s) of maximum bending moment could be obtained. However differentiating 1.12 gives 1.11, so by equating the shear force expression to zero the same result is obtained.

$$\therefore \quad wx - \frac{wl}{2} = 0$$

$$wx = \frac{wl}{2} \text{ so that } x = \frac{l}{2}$$

Although this method is not commonly used for determining bending moments and shear forces it has a useful application when the rate of loading is not constant. Example 1.8 illustrates this use.

Example 1.8
A beam 6 m long is simply supported at its ends and carries a load which varies uniformly in intensity from zero at one end to 5 kN/m at the other. Determine the magnitude of the maximum bending moment and where it occurs.

Referring to Fig. 1.10, the rate at which loading increases $= \frac{5}{6}$ kN/m/m. Thus, at a distance x from left-hand support;

$$w = \tfrac{5}{6} x \text{ kN/m}$$

$$\therefore \quad F = \int \tfrac{5}{6} x \, dx = \frac{5x^2}{12} + A. \text{ kN,}$$

and

$$M = \frac{5x^3}{36} + Ax + B. \text{ kN m}$$

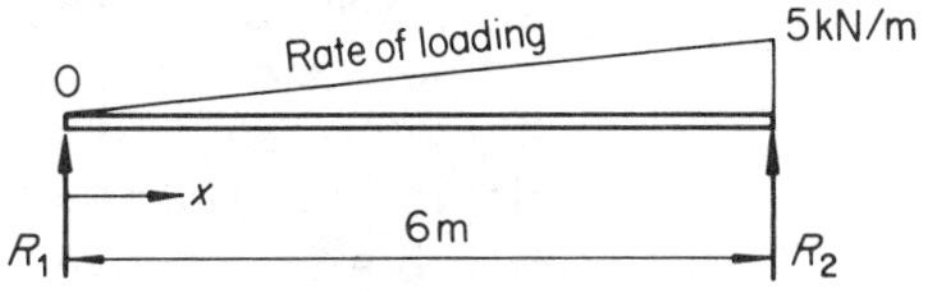

Fig. 1.10 A simply supported beam with a load varying uniformly in intensity

Inserting known conditions:

$M = 0$ when $x = 0$ $\therefore$ $B = 0$ and $M = 0$ when $x = 6$

$$\therefore \quad M = 0 = \frac{5 \times 6^3}{36} + 6A$$

$$A = -\frac{5 \times 6}{6} = -5 \text{ kN} = R_1$$

$$\therefore \quad F = \frac{5x^2}{12} - 5 \text{ kN} \tag{1.13}$$

$$\text{and} \quad M = \frac{5x^3}{36} - 5x \text{ kN m} \tag{1.14}$$

The maximum bending moment occurs when dM/dx or $F = 0$. Equating 1.13 to zero gives:

$$0 = \frac{5x^2}{12} - 5 \quad \text{or} \quad x^2 = \frac{60}{5} \quad \text{so that} \quad x = \pm 3.46 \text{ m}.$$

Therefore, from 1.14,

$$M_{MAX} = \frac{5 \times 3.46^3}{36} - 5 \times 3.46$$

$$= -11.55 \text{ kN m}.$$

The maximum bending moment is 11.55 kN m and occurs 3.46 m from the left-hand end.

1.9 Bending moment from shear force diagram

Referring once again to Equation 1.7, if $F \delta x = \delta M$, inserting limits and integrating gives:

$$\int_b^a F\,dx = \int_b^a \delta M$$

$$\text{or} \quad \int_b^a F\,dx = M_a - M_b$$

$\int_b^a F\,dx$ is the area of the shear force diagram between two points a and b. $M_a - M_b$ is the change in the bending moment between the same two points. Thus by starting at a section where the bending moment is zero it is possible to determine the bending moment at any other section by calculating the area of the shear force diagram between the two sections. Looking at the standard cases in Table 1.1 on page 6 this is clearly demonstrated.

(a) Simply supported beam, centre load

$$\text{BM at LH support} = 0$$

$$\text{BM at centre} = -\frac{Wl}{4}$$

and the difference between these $= -\frac{Wl}{4}$. The area of the shear force diagram from the LH support to the centre

$$= -\frac{W}{2} \times \frac{l}{2} = -\frac{Wl}{4}.$$

(b) Simply supported beam, distributed load

$$\text{BM at LH support} = 0$$

$$\text{BM at centre} = -\frac{wl^2}{8}$$

and the difference between these $= -wl^2/8$. The Area of the shear force diagram from the LH support to the centre

$$= -\frac{wl}{2} \times \frac{l}{4} = -\frac{wl^2}{8}.$$

1.10 Summary

(i) *Loading diagram*. Except for cantilevers, calculate the beam reactions. Sketch the diagram approximately or draw to scale.

(ii) *Shear force diagram*. Once all the forces, including the reactions, are known sketch the approximate diagram or draw using a suitable load scale.

1. With only concentrated loads, the SF remains constant between loads or reactions and changes abruptly at those loads or reactions.
2. With a distributed load only, the SF changes uniformly over the loaded section but abruptly at the support(s).
3. With mixed loading the SF changes uniformly between concentrated loads or reactions and abruptly at those loads or reactions.

(iii) *Bending moment diagram*. It is preferable to calculate the BM using the [ ] bracket notation. [ ] terms are only used when the value of the bracketted term is positive. Sketch the diagram approximately or draw it to scale.

1. For concentrated loads only, the BM changes uniformly between the loads and or reactions. The rate of change, or slope may vary between different loads.

2 For distributed loads only the BM varies with the square of the distance; the diagram is curved.
3 For mixed loads the diagram is curved over the parts of the beam where the distributed load is applied.

(iv) *Loading, shear force and bending moment*

$$F = \int w\, \delta x$$

$$M = \int F\, \delta x$$

$$M_a - M_b = \int_b^a F\, dx.$$

Exercises 1

Questions 1 to 5 refer to the theory and should be attempted initially without further reference to the text.

1 The following statements refer to a shear force diagram. Indicate which of them are correct.
  (a) The diagram indicates the change in bending moment between two sections.
  (b) From the diagram the section of maximum bending moment is determined.
  (c) It shows the variation in shear force from section to section.
  (d) Its area is related to the bending moment.
  (e) Its outline is curved for a uniformly distributed load.

2 With regard to the bending moment diagram for a loaded beam indicate which of the following statements are correct.
  (a) Bewteen points of constant bending moment the shear force is zero.
  (b) The diagram is curved where the loading is uniformly distributed.
  (c) A point of contraflexure occurs where the bending moment changes sign.
  (d) The bending moment is always zero at the ends of a beam.
  (e) The bending moment is always zero at the supports of a simply supported beam.

3 The rate of loading, the shear force and the bending moment are related. Which of the following relationships are correct?
  (a) $F = \int w\, \delta x.$

(b) $M = \int F \, \delta x$.

(c) $(M_a - M_b) = \int_b^a F \, \delta x$.

(d) The area of the loading diagram can be used to find the reactions.

(e) The area of the bending moment diagram can be used to find the shear force.

4 For a loaded beam the determination of the maximum bending moment is important. Indicate which of the following statements are correct, with regard to this, for a simply supported beam.
  (a) For a centrally loaded beam the maximum bending moment occurs at the centre.
  (b) The bending stress depends solely on the maximum bending moment.
  (c) The maximum bending moment occurs where the shear force changes sign.
  (d) For concentrated loads the maximum bending moment always occurs under a load.
  (e) The maximum bending moment is always at a section closer to the larger support reaction.

5 For a horizontal cantilever beam with both vertical concentrated and distributed loads:
  (a) the maximum shear force occurs at the support;
  (b) the maximum shear force is the algebraic sum of the loads;
  (c) the maximum bending moment occurs at the support;
  (d) the bending moment at any section is the total moment of all the forces between the section and the free end, about the section;
  (e) the support reaction is equal and opposite to the total moment on the beam.
Indicate which of these statements are correct.

6 A cantilever 2 m long carries an end load of 5 kN and a concentrated load of 10 kN at the mid point. Sketch the bending moment and shear force diagrams and insert significant values. Determine:
  (i) the bending moment at the mid point;
  (ii) the maximum bending moment and shear force.

7 A shaft projects 0.5 m from a rigid bearing and carries three gearwheels each of weight 400 N at distances of 0.5, 0.3 and 0.1 m from the bearing. Determine:
  (i) the maximum bending moment and shear force;
  (ii) the shear force and bending moment at the mid point.

continued

Sketch the shear force and bending moment diagrams inserting values at each load section.

8 A girder is used as a lifting gantry and projects 2 m from a wall. The girder has a weight of 1.5 kN/m. A load of 2 kN is supported at the mid point.
   Determine:
   (i) the shear force and bending moment at the load;
   (ii) the maximum shear force and bending moment;
   and display these values on a shear force and bending moment diagram sketched approximately to scale.

9 A cantilevered girder has a mass of 123.3 kg/m and is 0.75 m long. It carries concentrated loads of 2 kN, 1.5 kN and 3 kN at respective distances of 0.25, 0.5 and 0.75 m from the support. Determine, assuming $g = 9.81$ m/s^2:
   (i) the shear force and bending moment at each load;
   (ii) the maximum shear force and bending moment.
   Display the results on shear force and bending moment diagrams sketched approximately to scale. Curved or straight lines should be clearly distinguished.

10 A horizontal shaft is supported in spherical bearings over a span of 1.5 m. It carries gearwheels of weights 500 N, 400 N and 300 n at distances of 0.5, 0.8 and 1.2 m from the left-hand end. Determine:
   (i) the bearing reactions;
   (ii) the shear force and bending moment at each load section;
   (iii) the maximum bending moment and where it occurs.
   Sketch the bending moment and shear force diagrams.

11 A simply supported beam has a span of 2.7 m and carries concentrated loads of 5 kN 0.9 m from either end.
   Determine:
   (i) the shear force and bending moment at each load;
   (ii) the shear force and bending moment at the centre.
   Sketch the shear force and bending moment diagrams.

12 If the beam in question 11 had a weight equivalent to a distributed load of 2 kN/m repeat the solution.

13 A girder, which has a mass of 305.8 kg/m, is simply supported over a span of 2.5 m and carries concentrated loads of 4 kN, 6kN and 3 kN at distances of 0.5, 1.5 and 2 m respectively from the left-hand end. Sketch the load diagram and determine:
   (i) the reactions;
   (ii) the maximum bending moment and where it occurs;
   (iii) the maximum shear force.

Sketch the shear force and bending moment diagrams, inserting significant values. (Assume g = 9.81 m/s².)

14 A beam 3 m long is simply supported over a span of 2 m, one support being at the left-hand end. It carries a distributed load of 2 kN/m between the supports and concentrated loads of 1.5 kN and 4 kN at the free end at mid span, respectively. Determine the support reactions and sketch the shear force and bending moment diagrams inserting values at each support and under the loads.

15 For the beam loaded as described in question 14 write down an expression for the bending moment at any section, use the square bracket convention.
  (a) Derive the equation from:
    (i) the right-hand end;
    (ii) the left-hand end.
  (b) Using the equations derived in (a) determine the exact point of contraflexure.
  (c) Using the equation derived in (a) determine the bending moment 2.5 m from the right-hand end.

16 A beam is simply supported over a span of 0.75 m and carries a load which varies in intensity from zero at one end to 3 kN/m at the other.
  (a) Write down an expression for the loading.
  (b) Derive an expression for the shear force at any section.
  (c) Derive an expression for the bending moment at any section.
  (d) Determine the value and position of the maximum bending moment.

17 The weight of a simply supported beam is 3 kN/m. The span is 2.5 m and an additional load, which varies in intensity from 0 at the LHE to 2 kN/m at the RHE. Determine:
  (i) the support reactions;
  (ii) the position and value of the maximum bending moment.

18 The bending moment expression, derived from the left-hand end, for a simply supported beam is $-4.6x + 4[x-1] + 6[x-2]$
  (a) Sketch the loading diagram and show the support reactions.
  (b) Determine the position and value of the maximum bending moment.
Loads in kN.

2 Stress due to Bending

2.1 Introduction

The previous chapter dealt with the methods of determining the variation in bending moments and shearing forces for different load arrangements on components which acted as beams. It was thus possible to decide the greatest bending moment in each case and where it occurred. Since bending moments give rise to bending stresses, the maximum bending stress can also be determined. The calculation of bending stress is based on simple bending theory and it should be remembered that the stress is not uniform over a section but varies from a maximum tensile value, through zero to a maximum compressive value. As a reminder the Bending formula and the assumptions on which it is based are given in the next section.

2.2 Bending formula

Assumptions.

(i) Elastic properties are the same in tension and compression, i.e. $E_T = E_C$.
(ii) The beam is initially straight and symmetrical about the plane of bending.
(iii) The material obeys Hooke's Law.
(iv) The material is homogeneous and isotropic. (See Chapter 4 for a definition.)
(v) A right-angled transverse section before bending remains right-angled after bending.

Formula.

$$\frac{\sigma}{y} = \frac{M}{I} = \frac{E}{R}$$

M = Bending moment;
E = Modulus of elasticity;
R = Radius of curvature;
σ = Stress due to bending at distance y from neutral axis;
I = Second moment of area about the neutral axis.

In Fig. 2.1 the *neutral plane* and the *neutral axis* (NA) are shown. Along the neutral plane there is no strain and thus no stress. In the section view the neutral axis coincides with the neutral plane so that for the beam bent as shown, the layers above the neutral axis are in tension and the tensile stress will vary with the distance y_T from the neutral axis; y_T MAX occurs in the outer layer. Below the neutral axis the layers are in compression and the compressive stress varies with the distance y_c from the neutral axis; y_c MAX occurs in the bottom layer. If the beam is bent in the opposite direction (i.e. sagging) the stresses would obviously be reversed.

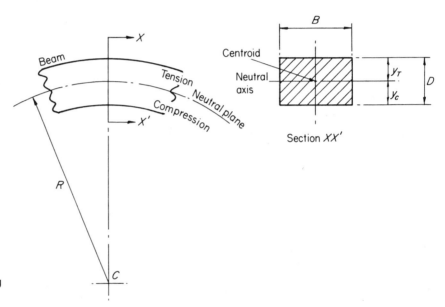

Fig. 2.1 Illustrating bending formula terms

For the beam section shown in Fig. 2.1 $I_{NA} = \dfrac{BD^3}{12}$.

The neutral axis is always assumed to pass through the centroid of the section. The radius of curvature, R, is that of the neutral plane and it is generally very large compared with the sectional dimensions.

Before considering a number of examples we must look at a method of determining the I values of some sections which would otherwise be more difficult to determine.

2.3 Parallel axis theorem

For certain beam sections the determination of the second moment of area is not straightforward if they are not rectangular or circular in shape. One method of overcoming this difficulty is to use the parallel axis theorem.

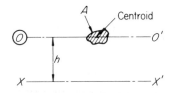

Fig. 2.2 A small area with an axis through its centroid

Fig 2.2 shows a small area, A, with an axis through its centroid OO'. The axis XX' is parallel to OO' but distance h from it. If I_{oo} is the second moment of area of the area A about axis OO' then its second moment of area about axis XX' is given by:

$$I_{xx} = I_{oo} + Ah^2$$

A simple demonstration of this is the rectangular section breadth B and depth D shown in Fig. 2.3. OO' is the axis through the centroid and $I_{oo} = \dfrac{BD^3}{12}$. XX' is an axis coincident with the base, and I_{xx} we know to be $\dfrac{BD^3}{3}$. However, using the parallel axis theorem $A = BD$ and $h = \dfrac{D}{2}$

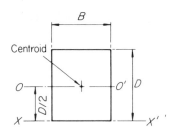

Fig. 2.3 A rectangular section

$$\therefore I_{xx} = \frac{BD^3}{12} + (BD)\frac{(D)^2}{(2)} = \frac{BD^3}{12} + \frac{BD^3}{4} = \frac{BD^3}{3}$$

Example 2.1
For the beam section shown in Fig. 2.4, determine the position of the centroid. Using the parallel axis theorem find the second moment of area of the section about a centroidal axis parallel to the base.

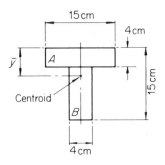

Fig. 2.4 The beam section

To find the position of the centroid, take moments about the top edge by dividing the section into two rectangles, A and B.

Then $\bar{y} = \dfrac{(15 \times 4 \times 2) + (11 \times 4 \times 9.5)}{(15 \times 4) + (11 \times 4)} = 5.173$ cm.

Then for rectangle A, $h = \bar{y} - 2 = 3.173$ cm;
$\qquad B, h = 9.5 - \bar{y} = 4.327$ cm.

Using the parallel axis theorem for each rectangle in turn:

$$I_{\text{CENTROID}} = \left\{\frac{15 \times 4^3}{12} + (15 \times 4 \times 3.173^2)\right\}$$
$$+ \left\{\frac{4 \times 11^3}{12} + (4 \times 11 \times 4.327^2)\right\}$$
$$= 1952 \text{ cm}^4$$

In the examples which follow, information from the worked examples in Chapter 1 is used in order to emphasise the importance of knowing the nature of the variation in bending moment so that the maximum bending stress can be determined.

Example 2.2
The shaft in Example 1.3, Chapter 1, is to be designed so that the maximum stress due to bending shall not exceed 30 MN/m². It is to be made of steel with a modulus of elasticity of 200 GN/m². Determine:

(a) the shaft diameter;
(b) the radius of curvature.

From $\dfrac{\sigma}{y} = \dfrac{M}{I}$, and 1.3:

$$M = BM_{MAX} = 56 \text{ Nm}$$

$$I = \dfrac{\pi d^4}{64} \text{ (for a circular section)}$$

$$y = d/2 \text{ (for a circular section)}$$

$$\sigma = 30 \times 10^6 \text{ N/m}^2.$$

$$\therefore \dfrac{I}{y} = \dfrac{\pi d^4}{64} \times \dfrac{2}{d} = \dfrac{M}{\sigma}$$

$$\therefore d^3 = \dfrac{32M}{\pi\sigma} = \dfrac{32 \times 56}{\pi \times 30 \times 10^6}$$

$$d = 0.0267 \text{ m or } 26.7 \text{ mm}.$$

From $\dfrac{E}{R} = \dfrac{\sigma}{y}$,

$$R = \dfrac{Ey}{\sigma} = \dfrac{200 \times 10^9 \times 0.01335}{30 \times 10^6}$$

$$R = 89 \text{ m}.$$

N.B. R is only approximate since the loading is not symmetrical. The fact that the shaft is bent into an arc due simply to the static weight of the gears will be significant in later work concerned with the whirling of shafts.

Example 2.3
The cantilever beam in Example 1.5, Chapter 1, is made of channel section which has cross-sectional dimensions as shown in Fig. 2.5. Determine the maximum stress due to bending and state where it occurs.

Fig. 2.5 (a) shows the section and Fig 2.5 (b) shows how it may be split into rectangles A, B and C, to which the parallel axis theorem can then be applied.

$$h \text{ for rectangles B and C} = 10.5 - 8.4 = 2.1 \text{ cm};$$

$$h \text{ for rectangle A} = 8.4 - 0.5 = 7.9 \text{ cm}.$$

Then $I_{CENT.} = 2\left\{\dfrac{1 \times 19^3}{12} + (1 \times 19 \times 2.1^2)\right\}$

$$+ \left\{\dfrac{10 \times 1^3}{12} + (10 \times 1 \times 7.9^2)\right\} = 1936 \text{ cm}^4$$

From Example 1.5 the maximum bending moment is 13.25 kN m and from Fig. 2.5 (a) the greater value of y is 0.116 m (11.6 cm) since the centroid, and thus the neutral axis, is in this case not equidistant from the top and bottom edges.

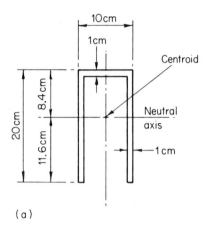

(a)

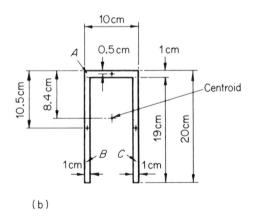

(b)

Fig. 2.5 The dimensioned cross-section of the channel section cantilever

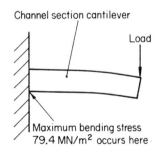

Fig. 2.6 Maximum bending stress

From $\dfrac{\sigma}{y} = \dfrac{M}{I}$ etc, $\sigma = \dfrac{My}{I} = \dfrac{13.25 \times 10^3 \times 0.116}{1.936 \times 10^{-5}}$ N/m²

$= 79.4 \times 10^6$ N/m² or 79.4 N/m².

Fig. 2.6 gives an exaggerated impression of the beam when loaded. The maximum bending moment occurs at the beam section where it enters the support. The maximum bending stress occurs along the bottom edge of this section i.e. at the base of the two vertical sides of the channel.

Example 2.4

The beam in Example 1.6, Chapter 1, is to have a maximum stress due to bending of 25 MN/m². Determine the section modulus of a suitable beam.

From the bending formula: $\dfrac{M}{I} = \dfrac{\sigma}{y}$.

The section modulus $(Z) = \dfrac{I}{y}$ so that $Z = \dfrac{M}{\sigma}$

The maximum bending moment from Fig. 1.7 is that which is numerically greater, i.e. 2.25 kN/m².

$$\therefore Z = \dfrac{2.25 \times 10^3}{25 \times 10^6} = 9 \times 10^{-5} \text{ m}^2$$

Since values of Z are usually quoted in cm³ this would be 90 cm³.

2.4 Eccentric loading

It is usually assumed that when direct tensile or compressive forces are applied to components the component axis and the load axis coincide. This means that true tension or compression can be assumed. Fig. 2.7 illustrates this for a compressive load. In certain practical situations such ideal conditions may not be present. The load axis may, for a number of reasons, be offset or eccentric from the component centroidal axis. This will lead to both bending and direct stresses being present and the resultant stresses will be greater than either of them.

Fig. 2.8 (a) and (b) shows a short column carrying a compressive load of W. The load axis is offset from the component axis by a distance e called the eccentricity. The cross-sectional dimensions of the column are shown in Fig. 2.8 (b).

$$\text{Direct compressive stress} = \dfrac{\text{load}}{\text{area}} = \dfrac{W}{bd}.$$

Bending moment due to eccentricity $= We$.

From the bending formula:

bending stress $(\sigma) = \dfrac{My}{I}$ where $M = We$

$y =$ distance from YY'

$$I = \dfrac{bd^3}{12}$$

$$\therefore \quad \sigma = \dfrac{Wey}{bd^3/12}$$

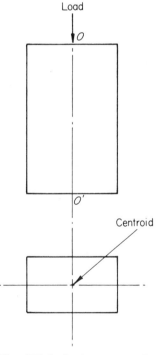

Fig. 2.7 A short component subject to a true compressive load—the load and centroidal axis coincide

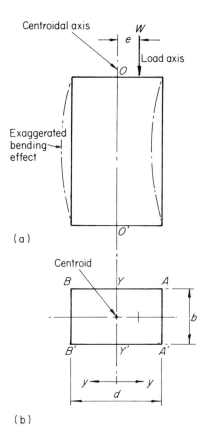

Fig. 2.8 (a) A short component subject to an eccentric load—the elevation shows bending due to eccentricity (b) Plan showing the dimensions of the section—YY' is neutral axis of bending

See ERRATA

when $y = d/2$,

$$\sigma_{MAX} = \frac{We\,d/2}{bd^3/12}$$

$$\therefore \sigma_{MAX} = \frac{6We}{bd^2}$$

This will be tensile along the edge BB' and compressive along the edge AA'. Along the axis YY' it will be zero.

Combining the direct compressive and the compressive bending stresses will give a maximum resultant compressive stress along edge $AA' = \sigma_{AA'} = \frac{W}{bd} + \frac{6We}{Bd^2}$. Along the edge BB' the resultant stress will be either tensile or compressive, according to whether the maximum tensile bending stress is greater or less than the direct compressive stress.

$$\therefore \sigma_{BB'} = \frac{W}{bd} - \frac{6We}{bd^2}.$$

For materials whose tensile strength is very much less than their compressive strength, the presence of a tensile stress when the applied load is compressive, is very significant. Examples of such materials are some of the cast metals and concrete in construction work. If the designer has arranged that components made of such materials are always subject to compressive stresses then the establishment of a tensile stress in practice by a non axial load may have disasterous consequences.

2.5 Eccentric loading: resultant stress distribution

Figure 2.9 (a)–(c) show respectively the compressive stress distribution, the bending stress distribution and the combined or resultant stress distribution when the maximum tensile bending stress is less than the direct compressive stress. In this case there is no resultant tension.

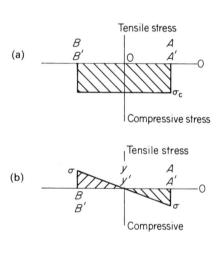

Fig. 2.9 Stress distribution with the bending stress greater than the compressive stress (a) Direct compressive stress (b) Bending stress (c) Resultant stress

Fig. 2.10 (a)–(c) show the respective distributions when the maximum tensile bending stress is greater than the direct compressive stress and resultant tension occurs.
At the section OO', where the resultant stress changes from compressive to tensile, there will be zero stress.

34 Mechanical Science IV

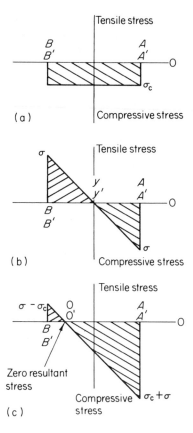

Fig. 2.10 Stress distribution with the bending stress smaller than the compressive stress (a) Direct compressive stress (b) Bending stress (c) Resultant stress

At section OO':

Direct compressive stress = tensile bending stress

$$\therefore \frac{W}{bd} = \frac{Wey}{bd^3/12} \text{ or } y = \frac{d^2}{12e} \tag{2.1}$$

If the resultant tension is to be avoided, then OO' must coincide with BB'. Thus:

$$y = d/2$$

∴ From equation 2.1,

$$\frac{d}{2} = \frac{d^2}{12e}$$

$$e = \frac{d}{6}$$

For the rectangular section, therefore, the eccentricity must not exceed $d/6$ if there is to be no resultant tension. Since the eccentricity could equally be on the opposite side of YY' a similar limit will apply. In general, the axis of zero stress is found from equating the direct and bending stresses:

$$\frac{W}{a} = \frac{Wey}{I_{NA}}$$

$$\text{or } y = \frac{I_{NA}}{ae} \qquad (2.2)$$

N.B. Since $I_{NA} = ak^2$, where a is the area and k the radius of gyration of the section about the neutral axis, Equation 2.2 could be written as $y = \frac{k^2}{e}$.

Example 2.5
A short concrete column of cross-section 0.25 m × 0.2 m carries a load of 4 kN acting at a distance of 0.1 m from the centroid on the longer axis. Determine the maximum resultant tensile and compressive stresses in the column if the applied load is compressive. Sketch the distribution diagram for the resultant stress indicating the section of zero stress.
What would the limiting eccentricity be if tension is to be avoided?

$$\text{Compressive stress} = \frac{\text{load}}{\text{area}} = \frac{4 \times 10^3}{0.25 \times 0.2}$$

$$= 80 \text{ kN/m}^2.$$

$$\text{Bending stress} = \frac{My}{I} \quad M = We = 4 \times 0.1$$

$$= 0.4 \text{ kN m}$$
$$= 400 \text{ N m}$$
$$y = 0.125 \text{ m}$$
$$I = \frac{0.2 \times 0.25^3}{12}$$
$$= 2.6 \times 10^{-4} \text{ m}^4$$

$$\therefore \frac{My}{I} = \frac{400 \times 0.125}{2.6 \times 10^{-4}}$$
$$= 192 \times 10^3 \text{ N/m}$$
$$= 192 \text{ kN/m}.$$

Maximum compressive stress = 80 + 192 = 272 kN/m².
Maximum tensile stress = 192 − 80 = 112 kN/m².
Axis of zero stress is obtained from Equation 2.2,

$$\therefore y = \frac{I_{NA}}{ae} = \frac{2.6 \times 10^{-4}}{0.25 \times 0.2 \times 0.1} = 0.052 \text{ m or } 5.2 \text{ cm from the neutral axis.}$$

To avoid tension the load must be within $d/6$ of the NA i.e.
$$\frac{0.25}{6} = 0.042 \text{ m}$$
$$= 4.2 \text{ cm}$$

See Fig. 2.11 (a) and (b) for the distribution diagram.
The theory is not limited to rectangular sections but it can be applied to any symmetrical section to which the simple bending theory applies.

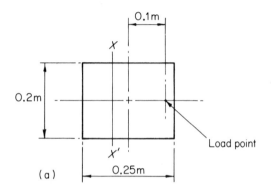

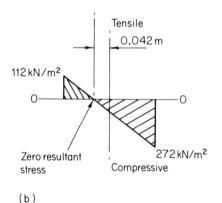

Fig. 2.11 (a) Plan view showing that the section of zero stress is YY'. (b) Stress distribution

Example 2.6
An *I* section column has both web and flanges 1.2 cm thick. The overall width of the flanges is 15 cm, and the overall depth of the section is 25 cm. The vertical compressive load on the column is 40 kN which is applied 9 cm from the centroid on the longer axis. Determine:
  (a) the maximum tensile and compressive resultant stress;
  (b) the section of zero stress;
  (c) Sketch the resultant stress distribution diagram.

Fig 2.12 (a) shows the arrangement of the load and the column section.

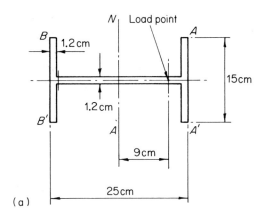

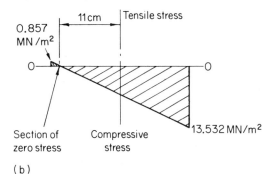

Fig. 2.12 (a) Plan view showing the dimensions and the load point. (b) The resultant stress distribution

Direct compressive stress $(\sigma_c) = \dfrac{\text{Load}}{\text{Area}} = \dfrac{W}{a}$.

$$\therefore W = 40 \times 10^3 \text{ N}$$
$$a = (15 \times 25) - (22.6 \times 13.8)$$
$$= 63.12 \text{ cm}^2$$
$$= 63.12 \times 10^{-4} \text{ m}^2$$

$$\therefore \sigma_c = \dfrac{40 \times 10^3}{63.12 \times 10^{-4}} = 6.34 \text{ MN/m}^2$$

Bending stress $(\sigma_B) = \dfrac{My}{I_{NA}}$

$$\therefore M = 40 \times 10^3 \times 0.09$$
$$= 3.6 \times 10^3 \text{ Nm}$$
$$y = \dfrac{0.25}{2} = 0.125 \text{ m}$$
$$I_{NA} = \left(\dfrac{15 \times 25^2}{12}\right) - \left(\dfrac{13.8 \times 22.6^3}{12}\right)$$
$$= 6257 \text{ cm}^4$$
$$= 6257 \times 10^{-8} \text{ m}^4$$

$$\therefore \sigma_B = \frac{3.6 \times 10^3 \times 0.125}{6257 \times 10^{-8}}$$
$$= 7.192 \text{ MN/m}^2.$$

$\therefore$ Maximum compressive stress along $AA' = 6.34 + 7.192$
$$= 13.532 \text{ MN/m}^2.$$
$\therefore$ Maximum tensile stress along $BB' = 7.192 - 6.34$
$$= 0.852 \text{ MN/m}^2.$$

From Equation 2.2, $y = \dfrac{I_{NA}}{ae} = \dfrac{6257 \times 10^{-8}}{63.12 \times 10^{-4} \times 0.09} = 0.110$ m.

$\therefore$ The section of zero stress is 11 cm from NA (See Fig. 2.12(b)).

2.6 Double eccentricity

If the load on the short column or component is eccentric from both the mutually perpendicular centroidal axes, as it may well be, then a second bending stress about the second axis will result. Referring to Fig. 2.13, if f is the second eccentricity then the bending stress about the axis ZZ' is given by:

$$\sigma_{zz} = \frac{Wfz}{db^3/12}.$$

z is the distance of any section from the axis ZZ' and $\dfrac{db^3}{12}$ is the second moment of area of the section about ZZ'.

The resultant stresses are now obtained from:

$$\sigma_R = \begin{array}{c}\text{direct} \\ \text{compressive} \\ \text{stress}\end{array} \pm \begin{array}{c}\text{bending} \\ \text{stress} \\ \text{about } YY'\end{array} \pm \begin{array}{c}\text{bending} \\ \text{stress} \\ \text{about } ZZ'\end{array}$$

$$= \frac{W}{a} \pm \frac{Wey}{I_{YY}} \pm \frac{Wfz}{I_{ZZ}} \qquad (2.3)$$

(Since $I = ak^2$, Equation 2.3 could be simplified to $\sigma_R = \dfrac{W}{a}\left(1 \pm \dfrac{ey}{k_{yy}^2} \pm \dfrac{f_z}{k_{zz}^2}\right)$ if desired. However it is the principle or method which is important so that in the worked examples the original form of 2.3 is used.)

Referring once again to Fig. 2.13 the resultant bending stresses in the four quarters of the section will be as follows (numbers corresponding to the quarters):

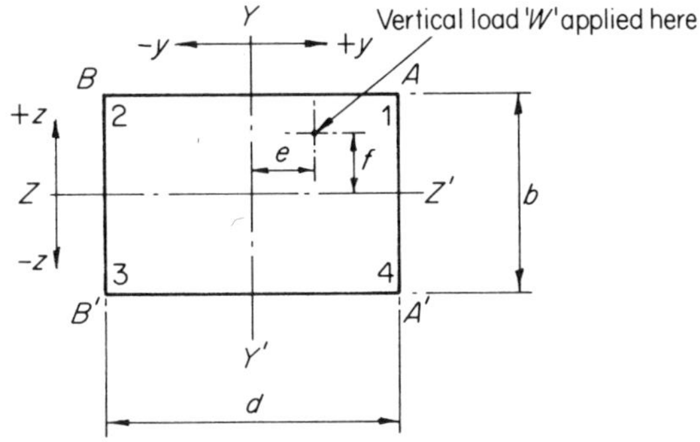

Fig. 2.13 Plan of a rectangular column showing double eccentricity

1 totally compressive;
2 compressive due to eccentricity f, and tensile due to eccentricity e;
3 both tensile, with the maximum total at corner B';
4 compressive due to eccentricity e and tensile due to eccentricity f.

To find the line of zero resultant stress, detect where this crosses the ZZ' or YY' axes.

When the resultant stress is zero on the ZZ' axis

$$\frac{W}{a} = \frac{Wey}{I_{YY}} \text{ since } \frac{Wfx}{I_{ZZ}} = 0 \text{ as } z = 0$$

When the resultant stress is zero on the YY' axis

$$\frac{W}{a} = \frac{Wfz}{I_{ZZ}} \text{ since } \frac{Wey}{I_{YY}} = 0 \text{ as } y = 0$$

Additional point on the line of zero stress are where it crosses the edges BB', $B'A'$ or AA', should there be no intersection with the ZZ' or YY' axes.

Example 2.7
A rectangular column of section 0.3 m × 0.2 m is subject to a vertical compressive load of 20 kN applied at a point 0.1 m from the shorter centroidal axis and 0.075 m from the other centroidal axis. Determine:

(a) the maximum bending and direct stress;
(b) the maximum resultant compressive and tensile stresses;
(c) the line of zero resultant stress.

Referring to Fig 2.14

$$I_{YY} = \frac{0.2 \times 0.3^3}{12} = 4.5 \times 10^{-4} \text{ m}^4$$

$$I_{ZZ} = \frac{0.3 \times 0.2^3}{12} = 2.0 \times 10^{-4} \text{ m}^4$$

Area (a) = 0.3 × 0.2 = 0.06 m²

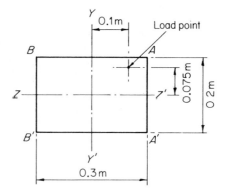

Fig. 2.14 Plan of column showing the dimensions and load point

Direct compressive stress $(\sigma_c) = \dfrac{W}{a} = \dfrac{20 \times 10^3}{0.06} = 333 \text{ kN/m}^2.$

Bending stress about YY' $\sigma_{YY} = \dfrac{Wey}{I_{YY}} = \dfrac{20 \times 10^3 \times 0.1 \times 0.15}{4.5 \times 10^{-4}}$

$= 666.7 \text{ kN/m}^2.$

Bending stress about ZZ' $\sigma_{ZZ} = \dfrac{Wfz}{I_{ZZ}} = \dfrac{20 \times 10^3 \times 0.075 \times 0.1}{2.0 \times 10^{-4}}$

$= 750 \text{ kN/m}^2.$

The maximum resultant compressive stress occurs at corner A

$= 333 + 666.7 + 750$
$= 1750 \text{ kN/m}^2.$

The maximum resultant tensile stress occurs at corner B'

$= -333 + 666.7 + 750$
$= 1.074 \text{ kN/m}^2.$

To find the line of zero resultant stress:

when $\sigma_{ZZ} = 0$ $y = \dfrac{d^2}{12e} = \dfrac{0.3^2}{12 \times 0.1} = 0.075 \text{ m}$

when $\sigma_{YY} = 0$ $z = \dfrac{0.2^2}{12 \times 0.075} = 0.044 \text{ m}.$

Joining these two points in Fig. 2.15 gives the line of zero stress. The additional point, P, on this line in Fig. 2.15 is obtained by equating equation 2.3 to zero.

$\therefore \sigma_R = 0 = \dfrac{W}{a} - \dfrac{Wey}{I_{YY}} + \dfrac{Wfz}{I_{ZZ}}$

compressive − tensile + compressive

$\dfrac{W}{a} = 333 \text{ kN/m}^2; \quad \dfrac{Wey}{I_{YY}} = 666.7 \text{ kN/m}^2$

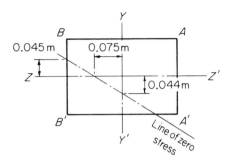

Fig. 2.15 The line of zero resultant stress

since $y = 0.15$ (maximum) so that only Z in $\dfrac{Wfz}{I_{zz}}$ is unknown

$$\therefore \sigma_R = 0 = 333 \times 10^3 - 666.7 \times 10^3 + \dfrac{20 \times 10^3 \times 0.075 \times z}{2 \times 10^{-4}}$$

$$\therefore z = 0.045 \text{ m}$$

2.7 Middle 'third' and middle 'quarter' rules

By considering the conditions for no tension due to eccentricities on each axis in turn, it is soon evident for a rectangular section that the shaded area in Fig. 2.16 shows the limits of any double eccentricity to ensure no resultant tension. The shaded area is within the middle third of the section.

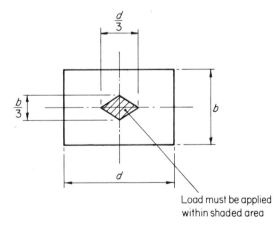

Fig. 2.16 Rectangular section showing the middle third rule for no tension due to eccentricity

For a circular section the critical area is a circle of diameter $d/4$, wher d is the section diameter. See Fig. 2.17.

2.8 Short columns and struts

It should be noted that the theory discussed in sections 2.4 to 2.7 will only apply to components whose lengths are comparable

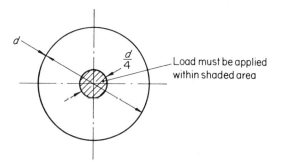

Fig. 2.17 Circular section showing the middle quarter rule for no tension due to eccentricity

with their cross-sectional dimensions. If the length is large compared with these dimensions then a different theory, the theory of struts must be applied. This theory is beyond the scope of this book, however it envisages that the longer component may buckle before failure occurs due to compression or tension.

2.9 Eccentric tensile loads

It is also possible for a tensile load to have the load axis displaced from the component axis. This will also result in bending stresses being set up in addition to the direct stress. Consequently, resultant compressive stresses may occur, although the applied load is tensile. The calculations are exactly the same as those for compressive eccentricity with the stresses reversed.

i.e. Resultant stress = direct tensile stress + tensile bending stress;

or resultant stress = direct tensile stress − compressive bending stress.

2.10 Summary

(i) *Bending formula* $= \dfrac{\sigma}{y} = \dfrac{M}{I} = \dfrac{E}{R}$.

(ii) *Bending stress due to eccentricity* $= \dfrac{Wey}{I_{YY}}$ or $\dfrac{Wfz}{I_{ZZ}}$.

(iii) *Resultant Stress due to eccentricity* $= \dfrac{W}{a} \pm \dfrac{Wey}{I_{YY}} \pm \dfrac{Wfz}{I_{ZZ}}$.

(iv) *Parallel Axis Theorem* — $I_{xx} = I_{oo} + Ah^2$.

Exercises 2

Questions 1 to 5 refer to the theory and should be attempted initially without further reference to the text.

1. The assumptions of the bending theory are included in the following statements. Indicate which are the correct statements.
   (a) $E_T = E_C$.
   (b) The section must be symmetrical about the plane of bending.
   (c) The material is of uniform structure.
   (d) The limit of proportionality is not exceeded.
   (e) The section must be symmetrical about the neutral axis of bending.

2. In connection with the bending theory, the neutral plane and the neutral axis are often referred to. The following statements concern them, indicate which are correct.
   (a) The neutral plane is unstrained during bending.
   (b) The neutral plane and the neutral axis are at right angles to each other.
   (c) The neutral plane is always midway between the top and bottom edges of the beam.
   (d) The neutral axis passes through the centroid of the section.
   (e) The bending stresses are always opposite on each side of the neutral axis.

3. With an eccentrically loaded component:
   (a) both bending and direct stresses are set up.
   (b) even if the load is compressive there will always be a resultant tensile stress due to bending.
   (c) the eccentricity is the distance from one edge of the section where the load is applied.
   (d) in compression, the theory applies only to components whose length is small compared with their cross-sectional dimensions.
   (e) the theory applies to eccentric tensile as well as to compressive loads.
   Indicate which of the above statements are correct.

4. When a compressive load is applied to a component with double eccentricity
   (a) it is assumed that the load axis is perpendicular to, but offset from, both the centroidal axes of the section.
   (b) no tensile resultant stress will occur if the load is applied to the middle third of a rectangular area.

continued

(c) no resultant tensile stress will result if the load is applied to the middle quarter of a circular area.
(d) the line of zero resultant stress is identified by the points where it cuts the neutral axes, or by the edges of the section.
(e) the resultant stress at the centroid is always zero.

Indicate the correct statements.

5 In connection with the application of the bending theory
(a) the maximum tensile and compressive bending stresses are always the same.
(b) the beam is always bent into a circular arc of a large but constant radius.
(c) the component must be initially straight.
(d) the load axis must be in the plane of bending.
(e) the I value is calculated relative to the neutral axis.

Indicate the correct statements.

6 A cantilever of rectangular section 400 mm deep and 300 mm wide carries a concentrated load of 20 kN at a distance of 1.5 m from the support. Determine:
   (i) the maximum stress due to bending;
   (ii) the radius of curvature if $E = 200 \text{ GN}/\text{m}^2$.

7 The elastic limit stress for a type of steel is 320 MN/m² and its modulus of elasticity is 207 GN/m². The steel is made into strip 4 mm thick and 0.25 m wide. It is coiled for the purpose of transport. Determine:
   (i) the minimum diameter of coil permissible without causing permanent bending;
   (ii) the maximum bending moment.

8 A beam is constructed of channel section with the open end pointing downwards. The external width of the channel is 150 mm and the depth 200 mm. The corresponding internal dimensions are 100 mm and 175 mm. Determine:
   (i) the position of the neutral axis relative to the top edge;
   (ii) the second moment of area about the neutral axis.

9 A beam is of I section of the following dimensions: the top flange is 20 cm × 5 cm deep; the web is 5 cm wide × 40 cm deep; the bottom flange is 30 cm wide × 5 cm deep. Determine:
   (i) the position of the neutral axis relative to the top edge;
   (ii) the second moment of area about the neutral axis.

10 A 25 mm diameter shaft is supported in spherical bearings over a span of 0.75 m. It is to carry two gearwheels each of weight 500 N which are to be placed 0.25 m from the respective bearings. If $E = 200 \text{ GN}/\text{m}^2$ determine:

(i) the maximum stress due to bending;
(ii) the radius of curvature of the shaft at the point of maximum stress.

11 A water pipe, which is 500 mm outside diameter and 400 mm inside diameter, is simply supported at two points 5 m apart. The pipe is full of water whose density is 1 000 kg/m³. The pipe material has a density of 7 Mg/m³. If $g = 9.81$ m/s² and $E = 200$ GN/m² determine;
(i) the maximum bending stress;
(ii) the radius of curvature, at the point of maximum stress.

12 A beam is constructed from a channel section and used as a cantilever with the open end pointing downwards. In this position the external width is 150 mm and depth 200 mm. The internal dimensions are 100 mm × 175 mm. The cantilever is used to support a motor whose weight is 4 kN. If the stress due to bending is not to exceed 30 MN/m², determine the maximum distance the motor may be placed from the support. The centroid of the channel section is 82.5 mm from the top edge.

13 A short concrete column of cross section 0.3 m × 0.2 m carries a vertical compressive load of 5 kN. The load is applied at a point 0.07 m from the centroid of the section along the shorter axis. Determine the resultant tensile and compressive stresses in the column and sketch the resultant stress distribution diagram for the section.

14 A short / section girder with flanges 300 mm wide and an overall depth of 400 mm is fixed at its lower end and projects vertically upwards. Both web and flanges are 25 mm thick. A component, whose weight is 2 kN, is attached to the outer edge of the flange so that its load axis on the flange centriline is vertical but 50 mm from the flange. Determine:
(i) the resultant maximum tensile and compressive stresses;
(ii) the section of zero stress relative to the outer edge.
(iii) Sketch the stress distribution diagram.

15 A short hollow circular section column projects vertically upwards from a concrete base. The outside diameter of the column is 200 mm and the inside diameter is 150 mm. A machine is clamped to the column by U-bolts so that its weight, of 4 kN, acts vertically on an axis x mm from that of the column. Determine the value of x if:
(i) the maximum compressive stress must not exceed 3 MN/m²;

continued

(ii) the maximum tensile stress must not exceed 3 MN/m².

16 A rectangular concrete column 0.4 m × 0.3 m carries a vertical compressive load of 30 kN. The load axis is on a diagonal of the column section 0.1 m from the centroid. Determine:
  (i) the maximum bending and direct stress;
  (ii) the maximum resultant tensile and compressive stresses;
  (iii) the line of zero resultant stress.

17 A vertical, short, cast-iron column has an outside diameter of 250 mm and an inside diameter of 200 mm. It carries a vertical load of 250 kN. The load eccentricity is 50 mm. Draw a stress distribution diagram showing how the stress varies across the section. Determine the eccentricity to just cause resultant tension.

18 A short, vertical column 0.2 × 0.3 m section carries a vertical load of 10 kN. The point of application of the load is 0.04 m from the shorter axis of the column section and 0.03 m from the longer axis. Determine the resultant maximum tensile and compressive stresses. Identify the line of zero stress of the section.

3 Deflection of Beams

3.1 Introduction

When a beam or other engineering component acting as a beam is subject to bending stresses, mechanical strain will result. This strain, like the stresses, may be both tensile and compressive at any section resulting in the beam being bent into an arc. Consequently parts of the beam will be deflected upwards or downwards relative to the support(s). For other than symmetrical loading the arc will not be of constant radius so that mathematical analysis of the curve becomes somewhat complex. However by using an approximated form of the analysis it is possible to extend the relationships established in Chapter 1 between loading, shear force and bending moment to include both the slope and deflection of the beam.

3.2 Beam curvature due to bending

Fig. 3.1 shows part of a beam subject to bending. The deflection and curvature are exaggerated for clarity. Considering the point A on the neutral plane which is distance x from one end and

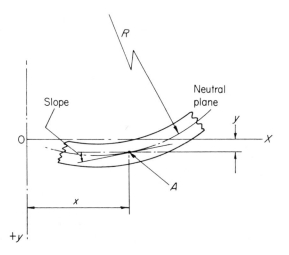

Fig. 3.1 The curvature, slope and deflection of a beam

deflected by an amount y from the unstrained position, if R is the radius of curvature at this point then:

$$\text{curvature} = \frac{1}{R} = \frac{d^2y}{dx^2}$$

This assumes that the slope at any point on the line of neutral plane is very small, as is the case with most beam calculations. From the bending formula:

$$\frac{M}{I} = \frac{E}{R} \quad \text{so that} \quad \frac{1}{R} = \frac{M}{EI} = \frac{d^2y}{dx^2}.$$

If this is integrated once we have:

$$\int \frac{d^2y}{dx^2} = \int \frac{M}{EI} dx \quad \text{or} \quad \frac{dy}{dx} = \frac{1}{EI} \int M\, dx;$$

and then a second time:

$$\int \frac{dy}{dx} = \frac{1}{EI} \int\int M\, dx\, dx \quad \text{or} \quad y = \frac{1}{EI} \int\int M\, dx\, dx.$$

Now dy/dx is the slope of the beam at the distance x from one end, and y is the deflection at the same point. If the bending moment, M, can be written as an equation in x then, by successive integrations, expressions for both slope and deflection are obtained. The method of writing the bending moment in this form was introduced in Chapter 1.

Two numerical examples are dealt with, initially, in order to emphasise the need, in most cases, to work from first principles. It is only when the maximum deflection alone is required that use of standard formulae is desirable.

Example 3.1
A cantilever beam carries an end load of 20 kN and has an EI constant of 40 kN m². The beam is 0.25 m long. Determine:
(a) the slope at the free end?
(b) the deflection at the free end;
(c) the slope at the mid point;
(d) the deflection at the mid point.

From Fig. 3.2 the bending moment at any distance x from the free end is given by:

$$M = 20x \text{ kN m}$$

$$\therefore \frac{d^2y}{dx^2} = \frac{M}{EI} = \frac{1}{EI}(20x);$$

$$\therefore EI \frac{d^2y}{dx^2} = 20x.$$

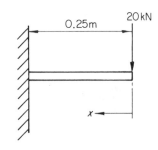

Fig. 3.2 The cantilevered beam with an end load

Deflection of Beams 49

The *EI* constant is transferred to the left-hand side of the equation in order not to confuse the integration.
Integrating once gives:

$$EI\frac{dy}{dx} = 20\frac{x^2}{2} + A = 10x^2 + A \qquad (3.1)$$

where A is the constant of integration. Integrating for a second time gives:

$$EIy = \frac{10}{3}x^3 + Ax + B \qquad (3.2)$$

and B is the constant of the second integration. The constants are evaluated by inserting known conditions. Thus the slope is zero where the beam enters the support.

∴ $dy/dx = 0$ when $x = 0.25$ and Equation 3.1 is equated to zero with $x = 0.25$

$$0 = 10 \times 0.25^2 + A$$
$$\therefore A = -0.625$$

Again when $x = 0.25$ the deflection $y = 0$, so that by equating 3.2 to zero:

$$EI \times 0 = \frac{10}{3} \times 0.25^3 - 0.625 \times 0.25 + B$$
$$0 = 0.052 - 0.156 + B$$
$$B = 0.104.$$

∴ Equation 3.1 becomes

$$\frac{dy}{dx} = \frac{1}{EI}(10x^2 - 0.625) \qquad (3.3)$$

equation 3.2 becomes

$$y = \frac{1}{EI}(3.33x^3 - 0.625x + 0.104). \qquad (3.4)$$

Before evaluation is carried out, the units of the bending moment expression and those of the *EI* constant should be made compatible. In this case they are already compatible, being kN m and kN m².

(i) In 3.3 put $x = 0$,

$$\therefore \frac{dy}{dx} = \frac{1}{40}(0 - 0.625) = -0.0156 \text{ Radians}$$

(ii) In 3.4 put $x = 0$,

$$\therefore y = \frac{1}{40}(0 - 0 + 0.104) = +2.6 \times 10^{-3} \text{ m}$$

$$\text{or } +2.6 \text{ mm}.$$

It should be noted that the slope is in *radians* and that the negative slope is the normal mathematical sense i.e. downwards from left to right. The deflection is positive, and with the sign convention used in the bending moment expression, positive deflections will be downward.

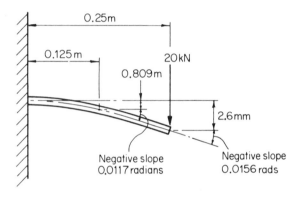

Fig. 3.3 (Note that the deflection and slope are very exaggerated)

(iii) In 3.3 put $x = 0.125$,

$$\therefore \frac{dy}{dx} = \frac{1}{40}(10 \times 0.125^2 - 0.625) = -0.0117 \text{ radians}$$

(iv) In 3.4 put $x = 0.125$,

$$\therefore y = \frac{1}{40}(3.33 \times 0.125^3 - 0.625 \times 0.125 + 0.104)$$

$$= +8.09 \times 10^{-4} \text{ m} = +0.809 \text{ mm}.$$

The answers are illustrated in Fig. 3.3.

Example 3.2
A length of shafting 200 mm diameter is supported in self aligning bearings (simply supported) which are 2 m apart. The shaft carries an evenly distributed load of 2 kN/m. If $E = 200$ GN/m², determine:
  (a) the deflection at the mid point;
  (b) the slope at the bearings.

From Fig. 3.4 by inspection $R_1 = R_2 = \frac{2 \times 2}{2} = 2$ kN. The bending moment expression is derived with x originating from the left-

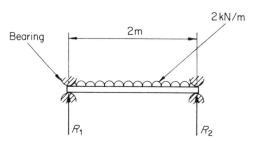

Fig. 3.4 The length of shafting supported in self aligning bearings

hand end, although it could equally well be from the right-hand end.

$$M = -R_1 x + \frac{wx^2}{2} = -2x + \frac{2x^2}{2}$$

$$\therefore EI\frac{d^2y}{dx^2} = -2x + x^2.$$

Integration gives:

$$EI\frac{dy}{dx} = -x^2 + \frac{x^3}{3} + A. \tag{3.5}$$

By integrating again,

$$EIy = -\frac{x^3}{3} + \frac{x^4}{12} + Ax + B. \tag{3.6}$$

To evaluate the constants it is known that:

(i) at the supports $x = 0$ or $x = 2$ and $y = 0$;
(ii) at the mid point $x = 1$ and $dy/dx = 0$.

Using Equation 3.6,

when $x = 0$ $y = 0$ $\therefore$ $EI \times 0 = 0 + 0 + 0 + B$ $\therefore$ $B = 0$;

when $x = 2$ $y = 0$ $\therefore$ $0 = \frac{2^3}{3} + \frac{2^4}{12} + 2A$ $\therefore A = +\frac{2}{3}.$

Alternatively, using Equation 3.5,

When $x = 1$ $dy/dx = 0$ $\therefore$ $EI \times 0 = -1^2 + \frac{1^3}{3} + A$

$$\therefore A = +\frac{2}{3}.$$

Equation 3.5 now becomes

$$\frac{dy}{dx} = \frac{1}{EI}\left(-x^2 + \frac{x^3}{3} + \frac{2}{3}\right). \tag{3.7}$$

Equation 3.6 now becomes

$$y = \frac{1}{EI}\left(-\frac{x^3}{3} + \frac{x^4}{12} + \frac{2}{3}x\right). \tag{3.8}$$

Before the deflection and the slope can be evaluated the EI constant must be calculated as it is not given. Once again the bending moment expression has units of kN m so that the EI constant must be in units of kN m² and m⁴ respectively.

$$E = 200 \text{ GN/m}^2 = 200 \times 10^9 \text{ N/m}^2 = 200 \times 10^6 \text{ kN/m}^2$$

$$I = \frac{\pi d^4}{64} = \frac{\pi \times 0.2^4}{64} = 7.854 \times 10^{-5} \text{ m}^4$$

$$\therefore EI = 200 \times 10^6 \times 7.854 \times 10^{-5} = 15708 \text{ kN m}^2$$

(i) From 3.8 the mid-point deflection is obtained by putting $x = 1$,

$$y = \frac{1}{15708}\left(\frac{1}{3} + \frac{1}{12} + \frac{2}{3}\right) = +2.65 \times 10^{-5} \text{ m}$$

$$= +2.65 \times 10^{-2} \text{ mm}.$$

(ii) From Equation 3.7 the slope at the ends is obtained by putting $x = 0$ or 2

$$\frac{dy}{dx} = \frac{1}{15708}\left(-0 + 0 + \frac{2}{3}\right) = +4.244 \times 10^{-5} \text{ rads}$$

or $$\frac{dy}{dx} = \frac{1}{15708}\left(-4 + \frac{8}{3} + \frac{2}{3}\right) = -4.244 \times 10^{-5} \text{ rads}.$$

The slopes have the same numerical value but are opposite in sign, as is to be expected. Fig. 3.5 illustrates the answer.

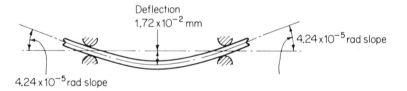

Fig. 3.5 (Note that the deflection and slope are very exaggerated)

Although the values of slope and deflection are extremely small they may nevertheless be significant in connection with vibrations and whirling. This topic is dealt with in Chapter 8.

The two examples have illustrated the method of dealing with simple cases of loading. Examples such as these are commonly encountered in engineering and are known as *standard cases*. For this reason they are dealt with in general terms and the results are tabulated and are often memorised. Four such cases will be dealt with in this section and a fifth will be covered in Section 3.4. The important results are tabulated in Table 3.1 in the summary.

3.3 Slope and deflection; standard cases

(a) Cantilever with concentrated end-load

Referring to Fig. 3.6(a) and (b).

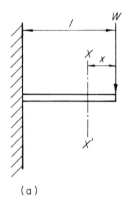
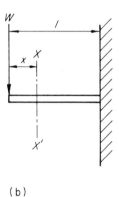

Fig. 3.6 A cantilever with a concentrated end load

(a)　　　　　　　　　　(b)

$$M_{xx} = Wx$$

$$\therefore EI\frac{d^2y}{dx^2} = Wx$$

$$EI\frac{dy}{dx} = \frac{Wx^2}{2} + A$$

$$EIy = \frac{Wx^3}{6} + Ax + B$$

When $x = l$ $\frac{dy}{dx} = 0$ and $y = 0$.

$$\therefore EI\frac{dy}{dx} = 0 = \frac{Wl^2}{2} + A \text{ so that } A = -\frac{Wl^2}{2}.$$

also, $EIy = 0 = \frac{Wl^3}{6} + Ax + B = \frac{Wl^3}{6} - \frac{Wl^3}{2} + B$ so that $B = \frac{Wl^3}{3}$.

$$\therefore \frac{dy}{dx} = \frac{1}{EI}\left(\frac{Wx^2}{2} - \frac{Wl^2}{2}\right), \quad (3.9)$$

and

$$y = \frac{1}{EI}\left(\frac{Wx^3}{6} - \frac{Wl^2 x}{2} + \frac{Wl^3}{3}\right). \quad (3.10)$$

Both dy/dx and y are a maximum at the free end i.e. $x = 0$.

54 Mechanical Science IV

Putting $x = 0$ in Equations 3.9 and 3.10:

$$\text{maximum slope} = -\frac{Wl^2}{2EI}; \qquad (3.11)$$

$$\text{maximum deflection} = \frac{Wl^3}{3EI}. \qquad (3.12)$$

(b) Cantilever with a uniformly distributed load

Referring to Fig. 3.7(a) and (b),

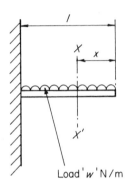

Fig. 3.7 A cantilever with a uniformly distributed load (a) (b)

$$M_{xx} = \frac{wx^2}{2}$$

$$\therefore EI\frac{d^2y}{dx^2} = \frac{wx^2}{2}$$

$$EI\frac{dy}{dx} = \frac{wx^3}{6} + A$$

$$EIy = \frac{wx^4}{24} + Ax + B.$$

When $x = l$ both dy/dx and $y = 0$.

$$\therefore EI\frac{dy}{dx} = 0 = \frac{wl^3}{6} + A \text{ so that } A = -\frac{wl^3}{6},$$

$$\text{and } EIy = 0 = \frac{wl^4}{24} + Ax + B = \frac{wl^4}{24} - \frac{wl^4}{6} + B$$

$$\text{so that } B = \frac{wl^4}{8}$$

$$\therefore \frac{dy}{dx} = \frac{1}{EI}\left(\frac{wx^3}{6} - \frac{wl^3}{6}\right) \qquad (3.13)$$

$$y = \frac{1}{EI}\left(\frac{wx^4}{24} - \frac{wl^3 x}{6} + \frac{wl^4}{8}\right) \quad (3.14)$$

Both dy/dx and y are a maximum at the free end i.e. $x = 0$.
Putting $x = 0$ in Equations 3.13 and 3.14:

$$\text{maximum slope} = -\frac{wl^3}{6EI}; \quad (3.15)$$

$$\text{maximum deflection} = \frac{wl^4}{8EI}. \quad (3.16)$$

(c) Simply supported beam with concentrated centre load

Referring to Fig. 3.8(a) and (b) between either end and the centre,

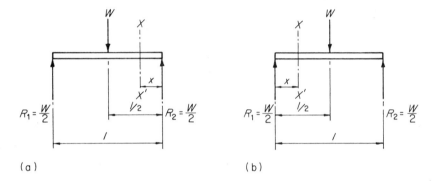

Fig. 3.8 A simply supported beam with a concentrated centre load

$$M_{xx'} = -\frac{Wx}{2}$$

$$\therefore EI\frac{d^2y}{dx^2} = -\frac{Wx}{2}$$

$$EI\frac{dy}{dx} = -\frac{Wx^2}{4} + A$$

$$EIy = -\frac{Wx^3}{12} + Ax + B$$

When $x = 0$ $y = 0$ thus $B = 0$.
When $x = \frac{l}{2}$, $\frac{dy}{dx} = 0$.

$$\therefore EI\frac{dy}{dx} = 0 = -\frac{Wl^2}{16} + A \quad \text{so that} \quad A = \frac{Wl^2}{16}$$

$$\therefore \frac{dy}{dx} = \frac{1}{EI}\left(-\frac{Wx^2}{4} + \frac{Wl^2}{16}\right); \quad (3.17)$$

and
$$y = \frac{1}{EI}\left(-\frac{Wx^3}{12} + \frac{Wl^2 x}{16}\right). \quad (3.18)$$

The maximum slope occurs when $x = 0$, i.e. at the supports; the maximum deflection occurs at the centre i.e. where $x = l/2$. Substituting these values in 3.17 and 3.18 respectively:

$$\text{maximum slope} = \pm\frac{Wl^2}{16 EI}; \quad (3.19)$$

$$\text{maximum deflection} = \frac{Wl^3}{48 EI}. \quad (3.20)$$

(d) Simply supported beam with a uniformly distributed load

Referring to Fig. 3.9(a) and (b) $R_1 = R_2 = \dfrac{wl}{2}$.

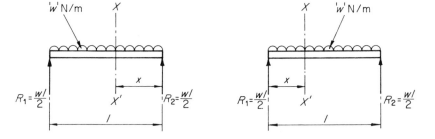

Fig. 3.9 A simply supported beam with a uniformly distributed load

$$M_{xx'} = -\frac{wlx}{2} + \frac{wx^2}{2}$$

$$\therefore EI\frac{dy}{dx} = -\frac{wlx^2}{4} + \frac{wx^3}{6} + A$$

and
$$EIy = -\frac{wlx^3}{12} + \frac{wx^4}{24} + Ax + B.$$

When $x = 0$ $y = 0$ thus $B = 0$.
When $x = l$ $y = 0$.

$$\therefore EIy = 0 = -\frac{wl^4}{12} + \frac{wl^4}{24} + Al \text{ so that } A = \frac{wl^3}{24}$$

$$\therefore \frac{dy}{dx} = \frac{1}{EI}\left(-\frac{wlx^2}{4} + \frac{wx^3}{6} + \frac{wl^3}{24}\right) \quad (3.21)$$

and
$$y = \frac{1}{EI}\left(-\frac{wlx^3}{12} + \frac{wx^4}{24} + \frac{wl^3 x}{24}\right) \quad (3.22)$$

The maximum slope occurs when $x = 0$ or l, at the supports. The maximum deflection occurs at the centre, i.e. where $x = l/2$.

Substituting these values in 3.21 and 3.22 respectively:

$$\text{maximum slope} = \pm \frac{wl^3}{24EI};$$

$$\text{maximum deflection} = \frac{5wl^4}{384EI}.$$

3.4 Macaulay's method

With the standard cases dealt with in the previous section, it was possible to write down a single expression for the bending moment at any section of the beams concerned, which was distance *x* from one end or the other. However, if a number of loads are present, it is not possible to derive a *single* expression for the bending moment, unless some artifice is used. This also applies if a single non-central load or a discontinuous distributed load is present.

It was suggested in Chapter 1 that by using square brackets, [ ], for terms which did not apply to the whole beam, the difficulty could be overcome providing these terms were only used when the value of the bracket was positive. This is the basis of Macaulay's method which further requires that the square brackets are integrated without expansion.

e.g. $$\int 3[x-4] = \frac{3}{2}[x-4]^2 + A$$

or $$\int \frac{3}{2}[x-4]^2 + A = \frac{3}{6}[x-4]^3 + Ax + B \text{ etc.}$$

The constants of integration, derived from known conditions, then apply to the whole beam.

The examples which follow illustrate the main features of the method. Note Examples 3.4 and 3.5 for particular types of distributed loads.

Example 3.3
A simply supported beam with a span of 10 m carries concentrated loads of 2 kN and 3 kN at the mid point and 2 m from the right-hand support, respectively. If $E = 200$ GN/m² and $I = 2 \times 10^{-5}$ m⁴ determine:

(a) the position and magnitude of the maximum deflection;
(b) the maximum slope.

Referring to Fig. 3.10 and taking moments about R_1,

$$10R_2 = (2 \times 5) + (3 \times 8)$$

$$R_2 = \frac{10 + 24}{10} = 3.4 \text{ kN};$$

$$R_1 = 5 - 3.4 = 1.6 \text{ kN}.$$

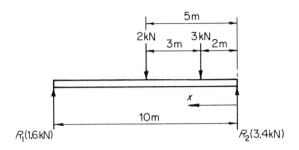

Fig. 3.10 A simply supported beam with two concentrated loads

Working from the right-hand end:

$$EI\frac{d^2y}{dx^2} = M_{xx'} = -3.4x + 3[x-2] + 2[x-5]$$

$$\therefore EI\frac{dy}{dx} = -1.7x^2 + \frac{3}{2}[x-2]^2 + \frac{2}{2}[x-5]^2 + A.$$

Remembering that the [ ] brackets remain intact for integrating,

$$EIy = -0.567x^3 + 0.5[x-2]^3 + \frac{1}{3}[x-5]^3 + Ax + B.$$

Since when $x = 0$ $y = 0$ then $B = 0$, the [ ] brackets are ignored.
When $x = 10$ $y = 0$ also, so that by substituting:

$$EIy = 0 = -567 + 0.5 \times 8^3 + \frac{1}{3} \times 5^3 + 10A$$

$$A = \frac{-567 + 256 + 41.67}{10}$$

$$= 26.9.$$

$$\therefore \frac{dy}{dx} = \frac{1}{EI}(-1.7x^2 + 1.5[x-2]^2 + [x-5]^2 + 26.9); \quad (3.25)$$

$$y = \frac{1}{EI}(-0.567x^3 + 0.5[x-2]^3 + 0.33[x-5]^3 + 26.9x). \quad (3.26)$$

To find the point of maximum deflection the slope is equated to zero, but in Equation 3.25 the dilemma is whether to include the [ ] terms or not. If the maximum deflection occurs at a section between R_2 and the 3 kN load, i.e. $x < 2$, then both the $[x-5]^2$ and the $1.5[x-2]^2$ terms are ignored. If the maximum deflection occurs between the loads then only the $[x-5]^2$ term is ignored since $x > 2$.
Obviously an initial 'guess' will have to be made and the result studied for its validity. With experience the correct guess can usually be made first time.

Guess 1. The maximum deflection occurs between R_2 and the 3 kN load i.e. $x < 2$. Equation 3.25 then becomes:

$$-1.7x^2 + 26.9 = 0 \quad \text{and} \quad x = \pm\sqrt{\frac{26.9}{1.7}} = \pm 3.98 \text{ m}.$$

This would appear to be valid until it is realised that the guess was that $x < 2$, so that a result $x = 3.98$ is not valid.
It is very important to remember this when looking at the solution.

Guess 2. The maximum deflection occurs between the 2 kN and 3 kN loads. i.e. $x > 2 < 5$.
Equation 3.25 then becomes:

$$-1.7x^2 + 1.5[x-2]^2 + 26.9 = 0$$

Expanding this:

$$-1.7x^2 + 1.5x^2 - 6x + 6 + 26.9 = 0$$

$$\therefore \quad -0.2x^2 - 6x + 32.9 = 0$$

$$x = \frac{6 \pm \sqrt{(36 + 26.32)}}{-0.4}$$

$$= -69.45 \quad \text{or} \quad +4.73 \text{ m}.$$

Note that the negative result is always disregarded.
Since $x = 4.73$ m (from the right-hand end) is within the section anticipated, then:

The maximum deflection occurs 4.73 m from the right-hand support.

From Equation 3.26:

$$x = 4.73 \text{ m} \quad \text{and} \quad EI = 200 \times 10^9 \times 2 \times 10^{-5}$$

$$= 4 \times 10^6 \text{ Nm}^2$$

The bending moment has units of kN m so that the EI constant must have units of kN m² i.e. $EI = 4 \times 10^3$ kN m².

$$\therefore \quad y_{max} = \frac{1}{4 \times 10^3}(-0.567 \times 4.73^3 + 0.5[4.73-2]^3$$

$$+ 26.9 \times 4.73)$$

$$= \frac{1}{4 \times 10^3}(-60 + 10.2 + 127.2)$$

$$= 0.019 \text{ m}$$

$$= 19 \text{ mm}.$$

The maximum slope occurs at the right-hand support as this is nearer to the maximum deflection. Thus, in Equation 3.25 we put $x = 0$, so that all the terms except the constants E, I and 26.9 disappear:

60 Mechanical Science IV

$$\frac{dy}{dx}_{max} = \frac{1}{4 \times 10^3} \times 26.9$$
$$= 6.725 \times 10^{-3} \text{ radians}$$
$$= 0.385°$$

Another difficulty which is encountered is that if a distributed load does not extend across the whole of the beam then the term representing this load in the bending moment expression may be invalid for certain sections. Examples 3.4 and 3.5 illustrate two methods by which this difficulty can be overcome.

Example 3.4
A beam 4 m long is simply supported at the left-hand end and at a point 3 m from the left-hand end. It carries a distributed load of 4 kN/m between the supports and a concentrated load of 6 kN at the right-hand end. Taking $E = 200$ GN/m² and $I = 3 \times 10^{-5}$ m⁴ determine:
(a) the deflection mid-way between the supports;
(b) the deflection and slope at the right-hand end.

Referring to Fig. 3.11 and taking moments about R_1:

$$3R_2 = (4 \times 3 \times 1.5) + (6 \times 4) = 42 \text{ kN m}$$
$$R_2 = 14 \text{ kN,}$$
$$R_1 = 12 + 6 - 14 = 4 \text{ kN.}$$

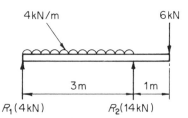

Fig. 3.11 Diagram showing the loading on the overhung beam

If the bending moment expression is derived from the left-hand end the terms for the distributed load would be $\frac{4x^2}{2}$. When $x > 3$ this would be invalid as the distributed load ceases at R_2. In this case, therefore, if we derive the bending moment expression $M_{xx'}$ from the right-hand end then the term for the distributed load is $\frac{4}{2}[x-1]^2$ which, if applied according to Macaulay rules, is valid for all sections.
∴ From the RHE

$$M_{xx'} = EI\frac{d^2y}{dx} = 6x - 14[x-1] + \frac{4}{2}[x-1]^2$$

$$EI\frac{dy}{dx} = 3x^2 - 7[x-1]^2 + \frac{2}{3}[x-1]^3 + A$$

$$EIy = x^3 - \frac{7}{3}[x-1]^3 + \frac{1}{6}[x-1]^4 + Ax + B$$

When $x = 1$ or 4, $y = 0$.
When $x = 1$, $EIy = 0 = 1^3 + A \times 1 + B$ ∴ $B = -(A+1)$.
When $x = 4$, $EIy = 0 = 64 - 63 + 13.5 + 4A - (A+1)$

$$A = -\frac{13.5}{3} = -4.5.$$

$$B = -(-4.5+1) = +3.5.$$

∴ $$\frac{dy}{dx} = \frac{1}{EI}\left(3x^2 - 7[x-1]^2 + \frac{2}{3}[x-1]^3 - 4.5\right), \quad (3.27)$$

and $$y = \frac{1}{EI}\left(x^3 - \frac{7}{3}[x-1]^3 + \frac{1}{6}[x-1]^4 - 4.5x + 3.5\right).$$

(3.28)

The EI constant is $(200 \times 10^9) \times (3 \times 10^{-5}) = 6 \times 10^6$ Nm²
$= 6 \times 10^3$ kN m².

(a) From Equation 3.28, $x = 2.5$ m, mid-way between the supports.

∴ $y = \frac{1}{6 \times 10^3}(15.625 - 7.875 + 0.844 - 11.25 + 3.5)$

$= \frac{+0.844}{6 \times 10^3} = +0.141 \times 10^{-3}$ m

$= +0.141$ mm

(b) From Equation 3.27, $x = 0$ at the RHE

∴ $\frac{dy}{dx} = \frac{1}{6 \times 10^3}(0 - 0 + 0 - 4.5)$

$= -7.5 \times 10^{-4}$ radians
$= -0.043°$

From Equation 3.28, $x = 0$ at the RHE

∴ $y = \frac{1}{6 \times 10^3}(0 - 0 + 0 - 0 + 3.5)$

$= +5.83 \times 10^{-4}$ m
$= 0.583$ mm.

Example 3.5
A shaft is simply supported in spherical bearings which are placed 2 m apart. The middle section carries a component whose weight

is equivalent to a distributed load extending 0.5 m either side of the mid point and equal to 4 kN/m. Another component positioned 1.5 m from the left-hand bearing is equivalent to a concentrated load of 8 kN. The *EI* constant is 3×10^3 kN m². Determine:
(a) the slope at the right-hand bearing;
(b) the deflection under the 8 kN load.

Referring to Fig. 3.12(a) and taking moments about R_1:
$$2R_2 = (4 \times 1 \times 1) + (8 \times 1.5)$$
$$\therefore R_2 = 8 \text{ kN}$$
$$R_1 = 4 \text{ kN}.$$

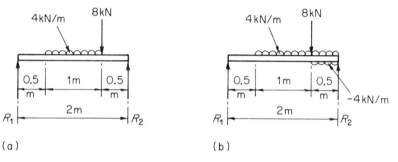

Fig. 3.12 (a) The shaft supported in spherical bearings with part distributed load (b) The shaft showing the modified loading

Since the distributed load does not extend to either end, the method used in Example 3.4 for dealing with its discontinuity cannot be used. In this case, therefore, the load is extended to one end, as shown in Fig. 3.12(b), and a negative load of 4 kN/m is applied over the extended portion to compensate. It is now possible, working from the left-hand end, to write down an expression for the bending moment which, applied according to the Macaulay rules, is valid for the whole beam.
From the LHE:

$$M_{xx'} = EI\frac{d_2 y}{dx^2} = -4x + \frac{4}{2}[x-0.5]^2$$
$$-\frac{4}{2}[x-1.5]^2 + 8[x-1.5]$$

$$EI\frac{dy}{dx} = -2x^2 + \frac{2}{3}[x-0.5]^3$$
$$-\frac{2}{3}[x-1.5]^3 + 4[x-1.5]^2 + A.$$

$$EIy = -\frac{2}{3}x^3 + \frac{1}{6}[x-0.5]^4$$
$$-\frac{1}{6}[x-1.5]^4 + \frac{4}{3}[x-1.5]^3 + Ax + B.$$

When $x = 0$, $y = 0$ so that $B = 0$.
When $x = 2$, $y = 0$ so that substituting:

$$EIy = 0 = -\frac{16}{3} + \frac{81}{96} - \frac{1}{96} + \frac{1}{6} + 2A$$

$$A = 2.17.$$

$$\therefore \frac{dy}{dx} = \frac{1}{EI}\left(-2x^2 + \frac{2}{3}[x-0.5]^3 - \frac{2}{3}[x-1.5]^3 \right.$$

$$\left. + 4[x-1.5]^2 + 2.17\right) \tag{3.29}$$

and $$y = \frac{1}{EI}\left(-\frac{2}{3}x^3 + \frac{1}{6}[x-0.5]^4 - \frac{1}{6}[x-1.5]^4 + \frac{4}{3}\right.$$

$$\left. \times [x-1.5]^3 + 2.17x\right). \tag{3.30}$$

(a) At the right-hand bearing $x = 2$ m.

∴ From Equation 3.29:

$$\frac{dy}{dx} = \frac{1}{3 \times 10^3}(-8 + 2.25 - 0.083 + 1 + 2.17)$$

$$= -8.87 \times 10^{-4} \text{ radians}$$

$$= 0.05°$$

(b) At the 8 kN load $x = 1.5$.

∴ From Equation 3.30:

$$y = \frac{1}{3 \times 10^3}(-2.25 + 0.167 + 3.255)$$

$$= 3.91 \times 10^{-4} \text{ m}$$

$$= 0.391 \text{ mm}.$$

3.5 Simply supported beam with a single, non-central, concentrated load

This is a further example of a commonly encountered form of loading, in addition to the standard cases dealt with in Section 3.3. It has been separated from those cases since it relies on an understanding of the Macaulay method.

Referring to Fig. 3.13:

$$R_1 = \frac{Wb}{l} \quad R_2 = \frac{Wa}{l} \quad \text{(see Chapter 1).}$$

Fig. 3.13 A simply supported beam with a single non-central, concentrated load

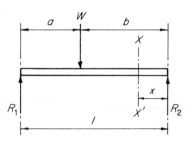

From the RHE, using Macaulay's method:

$$M_{xx'} = -\frac{Wa}{l}x + W[x-b].$$

$$\therefore EI\frac{d^2y}{dx^2} = -\frac{Wa}{l}x + W[x-b]$$

$$EI\frac{dy}{dx} = -\frac{Wa}{2l}x^2 + \frac{W}{2}[x-b]^2 + A$$

$$EIy = -\frac{Wa}{6l}x^3 + \frac{W}{6}[x-b]^3 + Ax + B$$

When $x = 0$, $y = 0$ so that $B = 0$.
When $x = l$, $y = 0$ so that substituting:

$$EIy = 0 = -\frac{Wa}{6l}l^3 + \frac{W}{6}[l-b]^3 + Al$$

But $[l-b] = a$,

$$\therefore 0 = -\frac{Wa}{6}l^2 + \frac{Wa^3}{6} + Al$$

$$A = \frac{Wa}{6l}l^2 - \frac{Wa^3}{6l}$$

$$= \frac{Wa}{6l}(l^2 - a^2).$$

Since $(l^2 - a^2) = (l-a)(l+a)$ and $(l-a) = b$,

then $A = \frac{Wab}{6l}(l+a)$.

$$\therefore \frac{dy}{dx} = \frac{1}{EI}\left\{-\frac{Wa}{2l}x^2 + \frac{W}{2}[x-b]^2 + \frac{Wab}{6l}(l+a)\right\}, \quad (3.31)$$

and

$$y = \frac{1}{EI}\left\{-\frac{Wa}{6l}x^3 + \frac{W}{6}[x-b]^3 + \frac{Wab}{6l}(l+a)x\right\}. \quad (3.32)$$

Under the load the deflection is given by putting $x = b$ in Equation 3.32.

The term $[x-b]^3$ is ignored since $x-b=0$ in this case.

$$\therefore \quad y = \frac{1}{EI}\left\{-\frac{Wab^3}{6l} + \frac{Wab^2}{6l}(l+a)\right\}$$

$$= \frac{1}{EI}\left\{\frac{Wab^2}{6l}(-b+l+a)\right\}$$

As $(-b+l+a) = 2a$,

$$y = \frac{Wa^2b^2}{3EIl}. \tag{3.33}$$

Equation 3.33 will be particularly useful in dealing with certain types of vibration in Chapter 8.

The maximum deflection occurs when $dy/dx = 0$ but in Equation 3.31 it is necessary to decide whether this is between R_2 and W, at W or between R_1 and W. Since $b > a$ the first option is true. The $[x-b]$ term will thus be ignored.

$$\therefore \quad EI\frac{dy}{dx} = 0 = -\frac{Wa}{2l}x^2 + \frac{Wab(l+a)}{6l}$$

$$\therefore \quad \frac{Wa}{2l}x^2 = \frac{Wab}{6l}(l+a)$$

$$x = \sqrt{\frac{b(l+a)}{3}}.$$

As $b = (l-a)$ and $(l-a)(l+a) = l^2 - a^2$, then

$$x = \sqrt{\frac{l^2 - a^2}{3}}. \tag{3.34}$$

The expression for the maximum deflection is not evaluated since in its general form it is not particularly useful.

3.6 Summary

(i) *Standard cases of slope and deflection of beams.* See Table 3.1
(ii) *Macaulay Method*

$$\frac{d^2y}{dx^2} = \frac{M_{xx}}{EI}$$

$$\frac{dy}{dx} = \frac{1}{EI}\int M_{xx} = slope$$

$$y = \frac{1}{EI}\int\int M_{xx} = deflection$$

Write down M_{xx} using [ ] brackets; use brackets when their value is positive; integrate bracket terms without expansion.
(iii) *General*
Apart from the standard cases, work from first principles.

Table 3.1.

Slope and deflection of beams: standard cases

	maximum slope	maximum deflection
(a) Cantilevers		
(i) concentrated end load	$\dfrac{Wl^2}{2EI}$ at free end	$\dfrac{Wl^3}{3EI}$ at support
(ii) distributed load	$\dfrac{wl^3}{6EI}$ at free end	$\dfrac{wl^4}{8EI}$ at support
(b) Simply supported beams		
(i) centre concentrated load	$\dfrac{Wl^2}{16EI}$ at supports	$\dfrac{Wl^3}{48EI}$ under load
(ii) distributed load	$\dfrac{wl^3}{24EI}$ at supports	$\dfrac{5wl^4}{384EI}$ at the centre
(iii) offset concentrated load	deflection under load $= \dfrac{Wa^2b^2}{3EIl}$	

Exercises 3

Questions 1 to 5 refer to the theory and should be attempted initially without further reference to the text.

1 In connection with the theory of the slope and deflection of beams the slope of the beam is:
   (a) always measured in radians.
   (b) the angle of the tangent to the curve of the beam at any point.
   (c) The angle at which the unloaded beam lies due to the position of the supports.
   (d) for most engineering beams, very small.
   (e) always, for any beam, has positive and negative values.
   Indicate the correct statements.

2 The curvature of a beam is:
   (a) equal to 1/radius of curvature.
   (b) in the theory used, always considered to be constant for the whole beam.
   (c) equals d^2y/dx^2 where y is the deflection and x the distance from one end.
   (d) equals M/EI in the bending theory.
   (e) expressed in mathematical terms.
   Indicate the correct statements.

3 The deflection of a beam:
   (a) is for a horizontal beam, the vertical displacement at any point due to the applied loads.
   (b) is related, mathematically, to the slope and curvature.
   (c) can be derived from the bending moment by a single integration.
   (d) depends upon the I value of the beam section.
   (e) will always be downwards.
   Indicate the correct statements.

4 The Macaulay method for finding the slope and deflection of a beam:
   (a) is based on the relationship between curvature and bending moment.
   (b) uses a single expression for the bending moment irrespective of the form of loading.
   (c) produces an expression for the deflection which applies, without modification, at all sections of the beam.
   (d) produces constants of integration which apply at all sections of the beam.
   (e) requires that, for some forms of loading, the section of maximum deflection must be roughly predicted.
   Indicate which statements are correct.

5 In connection with the slope and deflection calculations indicate which of the following statements are correct.
   (a) Square-bracketted terms are a feature of the Macaulay method.
   (b) Square-bracketted terms are expanded before integrating.
   (c) The points of maximum bending moment and deflection always coincide.
   (d) The slope is always zero at the maximum deflection.
   (e) It is possible, in some cases, to derive from the expression for the loading, an expression for the deflection.

continued

6 A cantilever which is 0.8 m long carries a single end load of 3 kN. If $E = 200$ GN/m² and $I = 128$ cm⁴ determine:
   (i) the slope at the free end;
   (ii) the deflection at the free end;
   (iii) the deflection at the mid-point.

7 A cantilevered beam carries a distributed load of 2 kN/m. Determine the greatest length if the deflection at the free end is not to exceed 5 mm.
$E = 200$ GN/m² and $I = 150$ cm⁴.
What would be the slope at the mid point?

8 A 15 mm diameter shaft is supported in spherical bearings 0.5 m apart. It carries a gearwheel of weight 180 N at (i) its mid point (ii) 0.2 m from one bearing. In each case determine the deflection of the shaft under the load. $E = 200$ GN/m².

9 A cantilevered gantry girder projects 2 m from its support. The maximum load to be carried is 10 kN. When the load is at the extremity of the gantry the deflection must not exceed 2 mm. Determine the necessary I value for the girder if $E = 200$ GN/m².

10 A large propellor shaft 0.3 m diameter is supported in spherical bearings over a span of 4 m. The shaft is made of steel of density 7 Mg/m³ and $E = 200$ GN/m². Determine the maximum deflection of the shaft.

11 A cantilever 0.5 m long carries an end load of 5 kN and a load of 3 kN 0.3 m from the support. Determine the maximum slope and deflection if the EI constant is 235 kN m² units.

12 A horizontal cantilever 3 m long carries a uniformly distributed load of 6 kN/m along its entire length and a concentrated load of 8 kN at its free end. The maximum deflection is limited to 50 mm. If $E = 200$ GN/m² determine the I value.

13 A beam is simply supported over a span of 1.5 m and carries two concentrated loads of 4 and 6 kN at distances of 0.3 and 1.0 m from the left-hand support respectively. If the EI constant is 25 kN m² determine:
   (i) the slope at the right-hand support;
   (ii) the deflection under each load.

14 A simply supported beam has a span of 2 m and carries concentrated loads of 6 kN and 5 kN at distances of 0.5 and 1.2 m from the left-hand support. The EI constant is 300 kN m² determine:
   (i) the maximum slope and where it occurs;
   (ii) the maximum deflection and where it occurs.

15 A shaft which is 2 m long is supported in spherical bearings. One bearing is at the left hand end and the other

1.5 m from that end. A wheel whose weight is 1.2 kN is fixed at the right hand end. A cylindrical component, fixed to the shaft, spans the distance between the bearings and has a weight equivalent to 3.2 kN/m. If the EI constant is 33 kN m² determine:
- (i) an expression for the slope at any section distance x from the left-hand end;
- (ii) an expression for the deflection at any section;
- (iii) the deflection mid way between the bearings;
- (iv) the slope at the right-hand end.

16 A simply supported girder has a span of 5 m and its weight is 1.5 kN/m. It carries a concentrated load of 10 kN, 2 m from the left-hand support. Determine:
- (i) the deflection under the concentrated load;
- (ii) the maximum slope and where it occurs.

$E = 200$ GN/m² and $I = 11\,000$ cm⁴.

4 Stress and Strain

4.1 Introduction

When a material is transmitting or resisting an applied force, it does so by becoming stressed. The nature of the stress is governed by the manner in which the force or load is applied, at least that is what the very simple stress theories imply. Thus, a tensile load (W) is associated with a tensile stress (σ) on a section XX' at right angles to the line of the load (Fig. 4.1). If a is the area of section XX', then:

$$\sigma = \frac{W}{a}$$

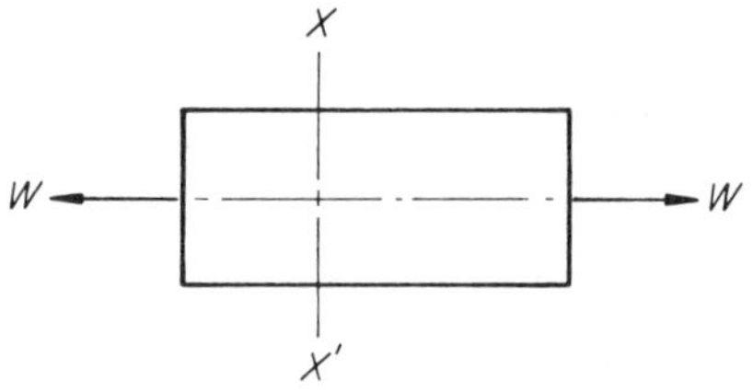

Fig. 4.1 Tensile load and stress

A shear load (W) is associated with a shear stress (τ) on a section XX' taken parallel to the load line (Fig. 4.2). If a is the area of section XX', then:

$$\tau = \frac{W}{a}$$

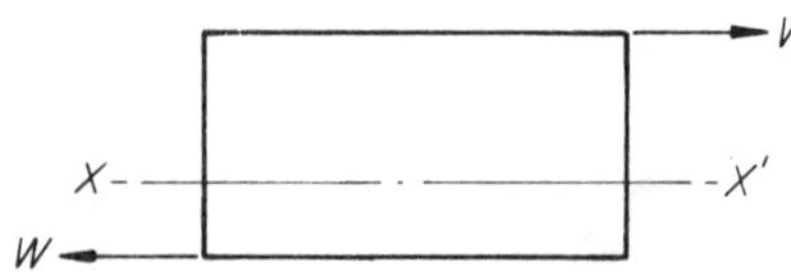

Fig. 4.2 Shear load and stress

What if we are now a little more inquisitive? Why, for instance, do some cast materials, when subject to a purely compressive load, fail in shear? Why, in the tensile testing of a ductile specimen of mild steel, does the cup and cone type of fracture occur? As shown in Fig. 4.3, the outer ring shows an obvious shear-type fracture.

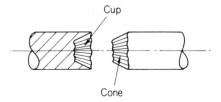

Fig. 4.3 Cup and cone fracture due to tension

The short answer to these questions is that tensile, compressive and shear stresses do not occur in isolation within a material. However, why this is so and how the total stress situation may be evaluated, needs now to be investigated.

4.2 Isotropic, orthotropic and non-isotropic materials

With the simple stress theory it was assumed that a material was homogeneous, which meant that its elastic properties were the same at all points and that the stress was uniform across the whole of a particular section. Now, however, we must begin to consider the three dimensions or planes in which stress may occur. In Fig. 4.4 dimensions XX', YY' and ZZ' illustrate these three dimensions.

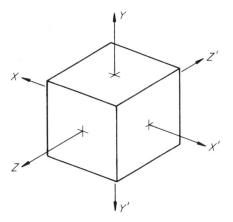

Fig. 4.4 The three dimensions in which stress may occur

A material which has the same elastic properties in each direction at all points is known as an isotropic material. Obviously, an isotropic material must also be homogeneous. Many of the common engineering materials are isotropic and it is this type of material alone which will be considered in the theory which follows. Referring to Fig. 4.4, a material may be homogeneous but still have different elastic properties in each of the three mutually perpendicular directions XX', YY', and ZZ', providing these properties are relatively the same at all points. Such a material is called orthotropic. Many engineering materials are processed, forged, rolled etc. to provide different properties in different directions. Wood, in its natural state is often a different strength across the grain compared to with the grain. These are all orthotropic materials.

If a material has different elastic properties from point to point it is said to be non-isotropic or alternatively, anisotropic, and it possesses many sets of elastic constants.

4.3 Poisson's ratio

A simple tensile load applied to a component produces a strain in the direction of the load which we call: *longitudinal strain*:

$$\text{longitudinal strain} = \frac{\text{change in length}}{\text{original length}} = \frac{x}{l}.$$

Experience will have revealed that the dimensions at right-angles to the load will also alter, as the necking, which occurs in a simple tensile test shows. The strain in these transverse dimensions is called: *lateral strain*:

$$\text{lateral strain} = \frac{\text{change in lateral dimension}}{\text{original lateral dimension}}$$

Experiments have shown that for engineering materials there is a relationship between the strains, within the elastic range of the material, which is referred to as *Poisson's ratio*:

$$\text{Poisson's ratio } (v) = \frac{\text{lateral strain}}{\text{longitudinal strain}}$$

Typical values for v are steel (carbon) 0.29; aluminium 0.34; brass 0.33; lead 0.45.

Example 4.1
Calculate the change in diameter of a pillar made of cast iron which carries a compressive load of 35 kN. The original diameter was 80 mm, the modulus of elasticity is 100 GN/m² and Poisson's ratio is 0.25.

$$\text{Longitudinal stress} = \frac{W}{a} = \frac{35 \times 10^3}{\pi/4 \times 80^2 \times 10^{-6}} = 6.96 \text{ MN/m}^2$$

From $E = \frac{\text{Stress } (\sigma)}{\text{Strain } (\varepsilon)}$

$$\varepsilon = \frac{-\sigma}{E} = -\frac{6.96 \times 10^6}{100 \times 10^9}$$

$$= -6.96 \times 10^{-5}$$

$$= -0.0696 \times 10^{-3}$$

$$\text{Lateral strain} = +v\left(\frac{\sigma}{E}\right) = 0.25 \times 0.0696 \times 10^{-3}$$

$$= 0.0174 \times 10^{-3}$$

Change in diameter = Original diameter × lateral strain
= 80 × 0.0174 × 10⁻³
= 1.39×10^{-3} mm.

The lateral strain is positive because due to the compressive load, the longitudinal strain is negative. The strains are always opposite in sign.

The previous example dealt with a circular section component for which there was only one lateral dimension i.e. the diameter. For rectangular sections there will be two lateral dimensions and the change in each may need to be determined. It is also possible for loads to be applied both longitudinally and laterally. In such cases it is necessary to distinguish between tensile and compressive strains. It is usual to call tensile strain positive (+) and compressive strain negative (−). It should be noted that tensile stress produces positive longitudinal strain but negative strains in both the lateral dimensions.

Example 4.2

Fig. 4.5 shows a rectangular sectioned component subject to two perpendicular loads. On the 100 mm × 50 mm face, there is a tensile load of 800 kN, and on the 200 mm × 50 mm face, there is a compressive load of 500 kN. If $E = 200$ GN/m² and Poisson's ratio is $\frac{1}{3}$, determine the total strains in each direction.

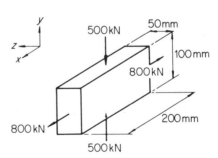

Fig. 4.5 The dimensions of a rectangular block

$$\text{Stress on the 100 mm} \times \text{50 mm face} = \frac{W}{a} = \frac{800 \times 10^3}{100 \times 50 \times 10^{-6}}$$
$$= 160 \text{ MN/m}^2 \text{ tensile.}$$

$$\text{Stress on the 200 mm} \times \text{50 mm face} = \frac{500 \times 10^3}{200 \times 50 \times 10^{-6}}$$
$$= 50 \text{ MN/m}^2 \text{ compressive.}$$

Stress on the 200 mm × 100 mm face = 0

Then from $E = \dfrac{\sigma}{\varepsilon}$:

in the direction of the 800 kN load, the total strain
ε_x = longitudinal strain
due to the 800 kN tensile load
+ lateral strain
due to the 500 kN compressive load.

$$\varepsilon_x = \frac{+160 \times 10^6}{200 \times 10^9} + \frac{1/3\,(50 \times 10^6)}{200 \times 10^9}$$
$$= 0.8 \times 10^{-3} + 0.083 \times 10^{-3}$$
$$= +0.883 \times 10^{-3} \text{ or } +0.000883;$$

in the direction of the 500 kN load, the total strain

$$\varepsilon_y = -\frac{50 \times 10^6}{200 \times 10^9} - \frac{1/3\,(160 \times 10^6)}{200 \times 10^9}$$
$$= -0.25 \times 10^{-3} - 0.267 \times 10^{-3}$$
$$= -0.517 \times 10^{-3} \text{ or } -0.000517;$$

in the direction z there is no direct load but the total strain is

$$\varepsilon_z = \frac{+1/3\,(50 \times 10^6)}{200 \times 10^9} - \frac{1/3\,(160 \times 10^6)}{200 \times 10^9}$$
$$= +0.083 \times 10^{-3} + 0.267 \times 10^{-3}$$
$$= +0.184 \times 10^{-3} \text{ or } +0.000184.$$

4.4 Volumetric strain

$$\text{Volumetric Strain } (\theta) = \frac{\text{change in volume}}{\text{original volume}} = \frac{\delta V}{V}.$$

Referring to Fig. 4.6 the volume of the block $V = xyz$.

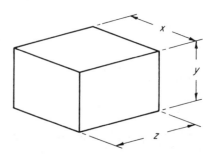

Fig. 4.6 The dimensions of the rectangular block

If the dimensions are increased by δx, δy, δz respectively then the new volume will be:

$$= (x+\delta x)(y+\delta y)(z+\delta z)$$
$$= (xy+\delta xy+\delta yx+\delta x\delta y)(z+\delta z)$$
$$= zxy+\delta xyz+\delta yxz+\delta x\delta yz+\delta zxy$$
$$+ \delta z\delta xy+\delta z\delta xx+\delta x\delta y\delta z.$$

Ignoring the products of small quantities i.e. $\delta x \delta y$, $\delta x \delta z$ etc,

$$\text{new volume} = xyz+\delta xyz+\delta yxz+\delta zxy;$$
$$\text{change in volume} = (xyz+\delta xyz+\delta yxz+\delta zxy) - xyz$$
$$= \delta xyz+\delta yxz+\delta zxy.$$

Dividing by the original volume xyz gives:

$$\text{volumetric strain } (\theta) = \frac{\delta xyz}{xyz}+\frac{\delta yxz}{xyz}+\frac{\delta zxy}{xyz}$$
$$= \frac{\delta x}{x}+\frac{\delta y}{y}+\frac{\delta z}{z}.$$

These are the three linear strains,

$$\therefore \text{ volumetric strain} = \text{Sum of linear strains.}$$

4.5 Stresses in thin cylinders and spheres

The volume of a cylinder will change when the diameter is strained; this is called *diametral strain*. The volume will also change if the length dimension is strained, i.e. *longitudinal strain*. Although some change in volume will occur if the thickness of the material is strained, this change can be ignored when the ratio of wall thickness to cylinder diameter is less than $\frac{1}{20}$. From this assumption we can develop the simplified theory applicable to stresses in thin cylinders and spheres.

Fig. 4.7 shows a thin cylinder of wall thickness t and diameter d. The cylinder is subject to an internal pressure p. This pressure will cause stresses to occur on circumferential sections, such as BB', and longitudinal sections, such as AA'.

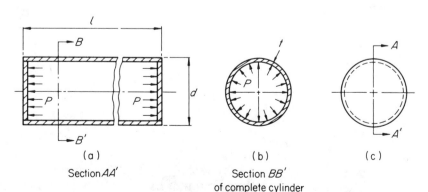

Fig. 4.7 A cylinder subject to an internal pressure p

(a) Section AA'

(b) Section BB' of complete cylinder

(c)

Hoop stress is due to the pressure acting radially, see Fig. 4.7(b), and it will set up the stress on longitudinal sections such as *AA'*.

$$\text{Hoop stress } (\sigma_H) = \frac{\text{force due to pressure}}{\text{area resisting force}}$$

$$= \frac{\text{pressure} \times \text{projected area}}{2 \times \text{length} \times \text{thickness}}$$

$$\sigma_H = \frac{pdl}{2tl} = \frac{pd}{2t}$$

Longitudinal stress is due to the pressure acting towards the ends and it will set up stress on circumferential sections such as *BB'*, see Fig. 4.7(a).

$$\text{Longitudinal stress } (\sigma_L) = \frac{\text{force due to pressure}}{\text{area resisting force}}$$

$$= \frac{\text{pressure} \times \text{end area}}{\pi dt}$$

$$\sigma_L = \frac{p\frac{\pi}{4}d^2}{\pi dt} = \frac{pd}{4t}$$

The hoop stress is thus double the longitudinal stress and it is the more significant of the stresses. Longitudinal joints therefore need to be stronger than circumferential joints.

For a spherical shell, only radial pressure occurs as shown in Fig. 4.8.

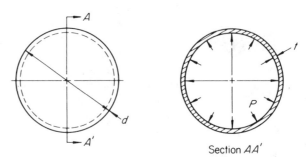

Fig. 4.8 A spherical shell subject to radial pressure

$$\text{Stress } (\sigma) = \frac{\text{pressure} \times \text{projected area}}{\text{area of circumferential section}}$$

$$\sigma = \frac{p\frac{\pi}{4}d^2}{\pi dt} = \frac{pd}{4t}$$

A spherical shell is thus stronger than a cylindrical one as the maximum stress is half that for a cylinder.

4.6 Change in volume of thin cylinders and spherical shells

Volumetric strain = sum of linear strains (Section 4.4.)
= 2 × hoop strain + longitudinal strain

Taking into account Poisson's ratio (v), Section 4.3,

$$\text{hoop strain} = \frac{\sigma_H}{E} - \frac{\sigma_L v}{E} = \frac{1}{E}(\sigma_H - v\sigma_L).$$

$$\text{longitudinal strain} = \frac{\sigma_L}{E} - \frac{v\sigma_H}{E} = \frac{1}{E}(\sigma_L - v\sigma_H).$$

For a thin cylinder,

$$\text{volumetric strain} = \frac{2}{E}(\sigma_H - v\sigma_L) + \frac{1}{E}(\sigma_L - v\sigma_H).$$

For a thin spherical shell,

$$\text{volumetric strain} = 3 \times \text{linear strain} = \frac{3}{E}(\sigma - v\sigma) = \frac{3\sigma}{E}(1 - v).$$

Example 4.3

An air pressure reservoir of cylindrical shape is made of steel 6 mm thick and it has a nominal diameter of 0.5 m. When it is subjected to an internal pressure of 10 bar determine:
 (a) the maximum stress;
 (b) the change in diameter;
 (c) the change in volume.
$E = 200 \text{ GN/m}^2$; $v = 0.3$, and the length of the cylinder is 2 m.

$$\text{Maximum stress} = \text{hoop stress} = \frac{pd}{2t} = \frac{10 \times 10^5 \times 0.5}{2 \times 6 \times 10^{-3}}$$

$$= 41.7 \text{ MN/m}^2$$

Since $\sigma_L = \sigma_H/2$,

$$\text{longitudinal stress} = \frac{41.7}{2} = 20.85 \text{ MN/m}^2$$

$$\text{diametral strain} = \frac{1}{E}(\sigma_H - v\sigma_L)$$

$$= \frac{10^6}{200 \times 10^9}(41.7 - 0.3 \times 20.85)$$

$$= 1.74 \times 10^{-4}$$

change in diameter = $0.5 \times 1.74 \times 10^{-4} = 0.87 \times 10^{-4}$ m or 0.087 mm.

$$\text{volumetric strain} = \frac{10^6}{200 \times 10^9}\{2(41.7 - 0.3 \times 20.85)$$

$$+ (20.85 - 0.3 \times 41.7)\}$$
$$= 3.96 \times 10^{-4}$$

original volume $= \dfrac{\pi}{4} \times 0.5^2 \times 2$

$= 0.393\,\text{m}^3$

change in volume $= 0.393 \times 3.96 \times 10^{-4}$
$= 1.56 \times 10^{-4}\,\text{m}^3 = 156\,\text{cm}^3$.

Example 4.4
A thin spherical pressure vessel of 6 m diameter is to be made out of steel plate. The maximum pressure is to be 10 bar and the maximum safe stress 90 MN/m². If $E = 200\,\text{GN/m}^2$ and $v = 0.3$ determine:
(a) the plate thickness;
(b) the change in volume at maximum pressure.

From $\sigma = \dfrac{pd}{4t}$, $t = \dfrac{pd}{4\sigma}$,

$$\therefore t = \dfrac{(10 \times 10^5)3}{4 \times 90 \times 10^6}$$

$= 8.33 \times 10^{-3}\,\text{m}$ or 8.33 mm.

Change in volume = original volume × volumetric strain

$$\therefore = \dfrac{4\pi}{3} \times 3^3 \times \dfrac{3 \times 90 \times 10^6}{200 \times 10^9}(1 - 0.3)$$

$= 0.107\,\text{m}^3$ or 107 cm³

4.7 Bulk modulus

If a material is subjected to equal stresses or pressures in each of the three dimensions, such as that produced most commonly by hydrostatic pressure, a change in volume will occur.

Then $\dfrac{\text{hydrostatic pressure (or equal 3-dimensional stress)}}{\text{volumetric strain}}$

$=$ *bulk modulus (K)*

i.e. $K = -\dfrac{P}{\partial V/V}$.

(The negative sign is used because the pressure or stress is compressive.)
The bulk modulus is related to the other elastic constants E, G and v; these relationships are developed in Section 4.12.

Example 4.5
A cube of material 150 mm × 150 mm × 150 mm is subjected to equal compressive three-dimensional stresses of 35 MN/m².

If $K = 70 \, \text{GN/m}^2$ determine the change in volume.

$$K = \frac{-P}{\delta V/V}$$

$$\frac{\delta V}{V} = -\frac{P}{K}$$

$$= \frac{35 \times 10^6}{70 \times 10^9}$$

$$= 5 \times 10^{-4}$$

$$\delta V = 5 \times 10^{-4} \times V$$

$$= 5 \times 10^{-4} \times 0.15^3$$

$$= 1.69 \times 10^{-6} \, \text{m}^3 \text{ or } 1690 \, \text{mm}^3$$

4.8 Complimentary shear stress

The element of material shown in Fig. 4.9 has dimensions x, y and z. Due to a shearing load a shear stress τ is set up on the top and bottom faces whose area is xz and which are marked A in Fig. 4.10. The total shearing force on each of these faces is then *stress × area* i.e. τxz. These two forces, acting at a distance apart of y will constitute a couple on the element of τxyz. If the material resists the shear force, then this element and all the other elements must remain in equilibrium. Since a couple can only be balanced

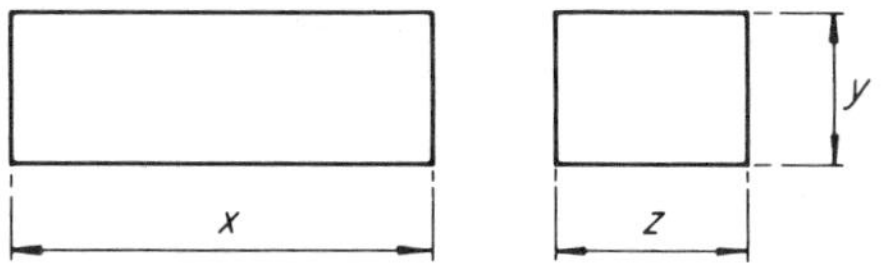

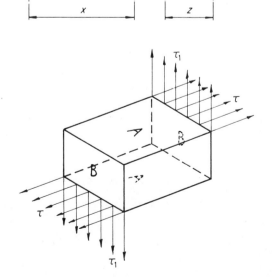

Fig. 4.9 The dimensioned element of material

Fig. 4.10 Diagram to show the couple effect of a shear stress

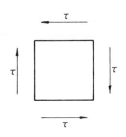

Fig. 4.11 Complimentary shear stress—pairs of stresses

by an equal and opposite couple, such a couple can only come from a shear stress acting on faces B in Fig. 4.10. If this stress is τ_1, since the area of the faces is yz and the distance between them is x, the couple is $\tau_1 xyz$.

Equating couples: $\tau xyz = \tau_1 xyz$

i.e. $\tau = \tau_1$

From this simple analysis we have established that a simple shear stress cannot and does not exist in isolation but it is *always* accompanied by an equal and opposite shear stress on a plane at right angles to the plane of the applied stress. The second stress is called the *complimentary shear stress*.

Fig. 4.11 shows that the pairs of stresses both act either towards or away from the corners of the element.

4.9 Complex stresses

(i) General

The previous section dealing with complimentary shear stress demonstrated quite clearly that a simple shear stress does not exist in isolation. Is it possible then that simple tensile or compressive stresses give rise to other stresses? The evidence mentioned in Section 4.1 suggested that this might be the case.

Using the technique suggested in Section 4.8, of considering a small element of material as being representative of all the material at a particular section, let us develop this idea a little further. Since such an element may be as small as we wish and have any dimensions we choose, then referring to Fig. 4.12 let the dimension $CC' = BB' = AA' = DD' = 1$.

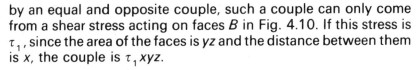

Fig. 4.12 The basic element

The area of face $DCC'D' = DC \times 1 = DC$

$ABB'A' = AB \times 1 = AB$

$BCC'B' = BC \times 1 = BC$

$ADD'A' = AD \times 1 = AD$

Similarly the area of an oblique section $ACC'A' = AC \times 1 = AC$.

If this is understood we may dispense with the 3-dimensional diagram of the element and use a simple elevation of the element, remembering that the depth is 1.

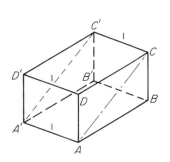

(ii) Simple tensile stress

Fig. 4.13 shows the element with a simple tensile stress acting on faces AB and DC. The *forces* exerted by this stress are:

$\sigma_1 \times$ AREA of DC and $\sigma_1 \times$ AREA of AB

i.e. $\sigma_1 DC$ and $\sigma_1 AB$.

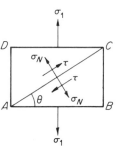

Fig. 4.13 The element with a simple tensile stress

Now let us investigate the forces acting on the oblique plane or section, AC. These forces must be in equilibrium with the forces acting on DC and AB. Considering the lower triangular piece of the element, ABC in Fig. 4.14(a) it is assumed for the moment that there are both normal and shear stresses on AC. It can only be from these stresses that the equilibrium forces are derived.

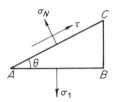
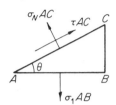

Fig. 4.14 (a) Stresses (b) Forces

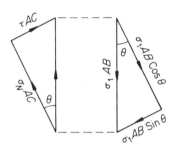

Fig. 4.15 (a) Applied force and components
(b) Resisting force and components

The first stage is to convert the stresses into forces by multiplying by the areas of the planes on which they are acting. This has been done in Fig. 4.14(b). In Fig. 4.15(a) the normal force on AB, $\sigma_1 AB$, has been resolved into two components namely $\sigma_1 AB \sin \theta$, parallel to AC, and $\sigma_1 AB \cos \theta$, normal to AC. In Fig. 4.15(b) the two forces assumed to be acting on face AC, namely $\sigma_N AC$ and τAC have been added vectorially. It should be apparent that the resultant is equal and opposite to $\sigma_1 AB$, the force on AB, and that individually

$$\tau AC = \sigma_1 AB \sin \theta \text{ and } \sigma_N AC = \sigma_1 AB \cos \theta.$$

Dividing by AC in each case, and noting that $\dfrac{AB}{AC} = \cos \theta$:

$$\tau = \sigma_1 \cos \theta \sin \theta \text{ and } \sigma_N = \sigma_1 \cos^2 \theta$$

$\therefore \sigma_N$ is a maximum when $\theta = 0°$ so that $\sigma_N = \sigma_1$.
This indicates that σ_1 is the greatest normal stress in the system. Later we shall define this as a *principal stress*. Note also that when $\theta = 0°$, $\tau = 0$. From $\tau = \sigma_1 \cos \theta \sin \theta$, and since $2 \cos \theta \sin \theta = \sin 2\theta$, we may write:

$$\tau = \frac{\sigma_1}{2} \sin 2\theta.$$

When $\theta = 45°$, τ is a maximum, i.e.

$$\tau_{max} = \frac{\sigma_1}{2}.$$

In other words our original simple tensile stress on AB, σ_1, gives rise to shear stresses on planes or sections oblique to AB. The maximum shear stress occurs on a plane at 45° to AB and is in magnitude equal to half the original tensile stress. Thus a material whose shear strength is less than half its tensile strength will, although loaded in tension, fail by shearing.

Example 4.6

A 50 mm diameter bar is subject to a tensile load of 50 kN. Determine the normal and tangential stresses on planes that make angles of (a) 30° and (b) 45° with the plane on which the normal tensile stress acts.

$$\text{Normal tensile stress} = \frac{\text{load}}{\text{area}} = \frac{50 \times 10^3}{\frac{\pi}{4} \times 0.05^2} = 25.5 \text{ MN/m}^2$$

On a plane at 30° to the normal stress plane: $\theta = 30°$; $2\theta = 60°$

$$\therefore \sigma_N = \sigma_1 \cos^2 30° = 25.5 \times 0.866^2 = 19.1 \text{ MN/m}^2$$

$$\tau = \frac{\sigma_1}{2} \sin 2\theta = \frac{25.5}{2} \times 0.866 = 11 \text{ MN/m}^2$$

On a plane at 45° to the normal stress plane: $\theta = 45°$; $2\theta = 90°$

$$\therefore \sigma_N = \sigma_1 \cos^2 45° = 25.5 \times 0.707^2 = 12.75 \text{ MN/m}^2$$

$$\tau = \frac{\sigma_1}{2} \sin 90° = \frac{25.5}{2} \times 1 = 12.75 \text{ MN/m}^2$$

(the maximum shear stress)

(iii) Simple compressive stress

If the original stress on plane AB of the basic element were compressive, see Fig. 4.16(a), then using the same analysis as in Section (ii)

$$\sigma_N = \sigma_1 \cos^2 \theta \text{ (compressive)}$$

$$\tau = \frac{\sigma_1}{2} \sin 2\theta \text{ (towards } A \text{ in Fig. 4.16(a).)}$$

Figs 4.16(b) & (c) illustrate the conversion of stresses into forces and the equating of the two components in each case.

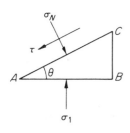
(a) Stresses

(b) Forces

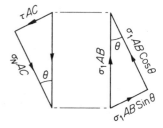
(c) Forces resolved

Fig. 4.16 The element with a simple compressive stress

Example 4.7

A prototype support column made of a cast material 150 mm diameter was subjected to a compression test to failure. The maximum load was found to be 4.5 MN but failure occurred by shearing along a plane at 40° to the axis of the column. Determine the compressive and the shear stress at fracture.

$$\text{Compressive stress at fracture} = \frac{\text{load}}{\text{area}} = \frac{4.5 \times 10^6}{\pi/4 \times 0.15^2}$$

$$= 255 \text{ MN/m}^2.$$

An angle of 40° to the column axis is 50° to the plane of the normal stress therefore, $\theta = 50°$ and $2\theta = 100°$:

$$\therefore \tau = \frac{\sigma_1}{2}\sin 2\theta = \frac{255}{2}\sin 100 = 125.6 \text{ MN/m}^2.$$

(iv) Simple shear stress

The thought should now occur that if simple tensile and compressive loads can give rise to shear stresses, is the reverse true and does a simple shear load result in tensile or compressive stresses being set up? When considering our basic element, initially loaded in shear on face AB (Fig. 4.17b), the complimentary shear on face BC must automatically be considered.

Fig. 4.17(a) shows the shear loading of the component and Fig. 4.17(b) shows the resultant shear stress on an element ABC coincident with the section XX' which is taken parallel to the load line. The complimentary shear stress has been included on face BC. It is now necessary to convert the stresses into forces by multiplying by the areas of the faces. These forces $\tau_1 AB$ and $\tau_1 BC$ are shown in Fig. 4.18(a) together with normal and tangential forces on the oblique face AC at an angle θ to AB. These forces are $\sigma_N AC$ and τAC respectively, their resultant is $\sigma_N AC$ at an angle ϕ to $\sigma_N AC$.

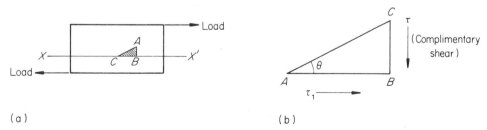

(a)   (b)

Fig. 4.17 (a) Shear loading (b) A magnified element of section XX' showing the applied and complimentary shear stresses

For equilibrium the applied forces $\tau_1 AB$ and $\tau_1 BC$ must be counterbalanced by the normal and tangential forces on the face AC or their resultant. Figs 4.18(b) and (c) show the applied forces resolved into components normal and tangential to AC. Fig. 4.18(d) shows the vector sum of $\sigma_N AC$ and τAC to give their resultant $\sigma_N AC$.

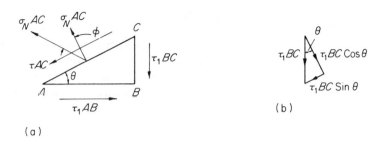

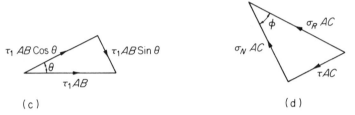

Fig. 4.18 (a) Forces on the element (b) Force on BC resolved (c) Force on AB resolved (d) Resisting forces on AC and the resultant force

Dealing first with the forces tangential or parallel to AC:

$$\tau AC = \tau_1 BC \sin \theta - \tau_1 AB \cos \theta.$$

Divide by AC noting that $\dfrac{BC}{AC} = \sin \theta$ $\dfrac{AB}{AC} = \cos \theta$:

$$\tau = \tau_1 \sin^2 \theta - \tau_1 \cos^2 \theta \qquad (4.1)$$

Then dealing with the forces normal to AC:

$$\sigma_N AC = \tau_1 BC \cos \theta + \tau_1 AB \sin \theta$$

Divide by AC as before:

$$\sigma_N = \tau_1 \sin \theta \cos \theta + \tau_1 \cos \theta \sin \theta$$
$$\sigma_N = \tau_1 2 \sin \theta \cos \theta$$
$$\sigma_N = \tau_1 \sin 2\theta \qquad (4.2)$$

From 4.2 when $\theta = 45°$,

$\sigma_N = \tau_1$ and is a maximum tensile

when $\theta = 135°$

$\sigma_N = -\tau_1$ and is a maximum but compressive

From 4.1 when $\theta = 45°$ or $135°$ $\tau = 0$.

The simple shear stress thus results in a normal stress on an oblique plane or face such as AC which reaches a maximum value when the face is at 45° or 135° to the plane of the original stress. The magnitude of these maximum normal stresses is equal to the magnitude of the original shear stress one being tensile and one being compressive.

Example 4.8
Across a certain section in a component there is a shear stress of 12 MN/m². Determine, for a section at 30° to the original section,

the normal, shear and resultant stress. Calculate the angle between the resultant stress and the normal stress. State the maximum tensile and compressive stresses set up and show, by a diagram, their relationship to the original stress.

From Equation 4.2:

$$\sigma_N = \tau_1 \sin 2\theta.$$

$\tau_1 = 12 \, \text{MN/m}^2; \, \theta = 30°; \, 2\theta = 60°.$

$$\sigma_N = 12 \sin 60° = 10.4 \, \text{MN/m}^2.$$

From 4.1,

$$\tau = \tau_1 \sin^2 \theta - \tau_1 \cos^2 \theta$$
$$= 12 \sin^2 30 - 12 \cos^2 30$$
$$= 3 - 9.0 = -6 \, \text{MN/m}^2$$

The negative sign for the shear stress means that it acts from A to C and not C to A as was assumed when Equation 4.1 was derived.

Resultant $= \sqrt{6^2 + 10.4^2} = 12.0 \, \text{MN/m}^2$

$$\phi = \tan^{-1} \frac{6}{10.4} = 30°$$

The maximum tensile and compressive stresses are $12 \, \text{MN/m}^2$ which are shown in Fig. 4.19.

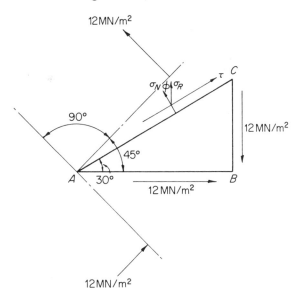

Fig. 4.19 The maximum tensile and compressive stresses

(v) Coplanar stress system

The investigation of simple shear, in the previous section, has shown that it is possible to analyse a stress system where there are stresses acting on two planes at right angles. In concentrating on this coplanar system it must not be overlooked that there may also be stresses in the third dimension. However since the coplanar system is the more common we will confine our attention to it, at this stage. Two simple examples of the system of stresses are a thin walled cylinder subject to longitudinal and hoop stress due to internal pressure, and a shaft transmitting power which is also subject to an end thrust resulting in both shear and compressive stresses.

Example 4.9

A cantilever beam 0.5 m long supports a concentrated end-load of 24 kN. The beam cross section is 9 cm wide by 12 cm deep. At a section coincident with the support determine, for a point 4 cm above the neutral axis, the normal and shear stresses and display them on an element. (N.B. it may be assumed that at this point the actual shear stress is equal to the mean shear stress on the whole section.)

$$\text{Mean shear stress} = \frac{\text{load}}{\text{area}} = \frac{24 \times 10^3}{12 \times 9 \times 10^{-4}} = 2.22 \text{ MN/m}^2.$$

$$\text{Bending stress } \sigma = \frac{My}{I} \quad M = 24 \times 10^3 \times 0.5 = 12 \times 10^3 \text{ N m}$$

$$y = 4 \times 10^{-2} \text{ m}$$

$$I = \frac{bd^3}{12} = \frac{9 \times 12^3}{12} \times 10^{-8}$$

$$= 1.296 \times 10^{-5} \text{ m}^4.$$

$$\sigma = \frac{12 \times 10^3 \times 4 \times 10^{-2}}{1.296 \times 10^{-5}}$$

$$= 37 \text{ MN/m}^2 \text{ Tensile.}$$

Fig. 4.20(a) shows the loaded beam; Fig. 4.20(b) the enlarged section and the point for which the stress is calculated; Fig. 4.20(c) shows the element whose face BC is part of the section.

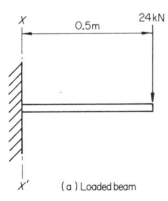
(a) Loaded beam

Fig. 4.20 (a) Loaded beam (b) Section XX' (c) Stresses on an element located at point P, the maximum shear stress

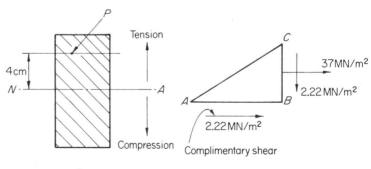

(b) Section XX'

(c) Stresses on an element located at point P

Stress and Strain 87

From this stress system it would be necessary to determine the maximum resultant stresses. The method of doing this is explained in the following sections.

4.10 Principal stresses and maximum shear stress

It should by now be very apparent that stress systems are much more complex than simple theory would suggest. It is necessary in all cases to be aware of the combined effects of tension, torsion and shear and to be able to predict the maximum resultant stresses.

(i) Principal stress

This is the name given to both the *maximum* and *minimum normal* resultant *stresses* derived from a stress system. The plane on which a principal stress acts is called a *principal plane* and it will be found that it is a *plane of zero shear stress*. Principal planes are 90° apart.

Although it is not intended to formally prove the above statements, examination of the equations derived in Section 4.9 (ii), (iii) and (iv) will indicate their correctness.

$$\text{From } \sigma_N = \sigma_1 \cos^2 \theta \text{ and } \tau = \frac{\sigma_1}{2} \sin 2\theta$$

When $\theta = 0°$ or $90°$ or $2\theta = 0°$ or $180°$

$\sigma_N = \sigma_1$ maximum value

or $\sigma_N = 0$ minimum value

$\tau = 0$ in both cases since $\sin 0° = 0$
$\sin 180° = 0$

From Equation 4.2: $\sigma_N = \tau_1 \sin 2\theta$;

and Equation 4.1: $\tau = \tau_1 \sin^2 \theta - \tau_1 \cos^2 \theta$.

when $\theta = 45°$ or $135°$,

$\sigma_N = \tau_1$ *tensile* or $-\tau_1$ *compressive*

i.e. the maximum and minimum values of σ_N
In both cases, with $\theta = 45°$ or $135°$, $\tau = 0$.

(ii) Maximum shear stress

This is the maximum value of the resultant shear stress in a system.

Maximum shear stress = $\frac{1}{2}$ *difference between the principal stresses*

and occurs on planes at 45° to the principal planes.
All these facts are summarised in Fig. 4.21.

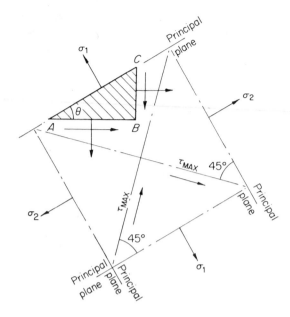

Fig. 4.21 The principal planes and stresses and the maximum shear stress

4.11 The Mohr stress circle

It is possible to use the kind of analysis developed in Section 4.9 to determine the principal stresses and maximum shear stress for a complete coplanar system. However the method is rather lengthy and a much quicker way is to use the Mohr stress circle which can give a purely graphical solution, with limited accuracy, or a calculated solution with the same degree of accuracy as the more analytical method. Fig. 4.22, (a) and (b), show respectively the various stresses applied to the basic element and their representation in the Mohr Circle.

From the starting or *pole point*, P,:
    tensile stresses are drawn horizontally to scale to the right.
    compressive stresses are drawn horizontally to scale to the left.
    positive shear stresses are drawn to scale vertically upwards, to scale.
    negative shear stresses are drawn vertically downwards, to scale,

(A positive shear stress is one that tends to give a clockwise rotation to the element, i.e. that on plane BC in Fig. 4.22(a) is positive.)

Referring to Fig. 4.22(a) and (b):
    $PN = \sigma_y$ the tensile stress on plane AB.
    $NR = \tau$, the negative shear stress on plane AB.
    $PR =$ the resultant stress on plane AB.
    $OR =$ the position of the plane AB in the stress circle.
    $PN' = \sigma_x$, the tensile stress on plane BC.

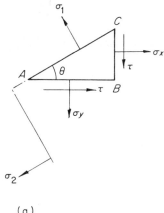

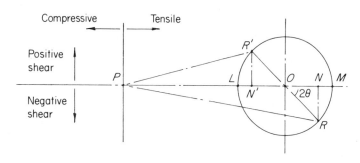

Fig. 4.22 (a) The element
(b) The Mohr stress circle

$N'R' = \tau$, the positive shear stress on plane BC.
$PR' = $ the resultant stress on plane BC.
$OR' = $ the position of plane BC in the stress circle.
$PL = $ the principal stress, σ_2.
$OL = $ the position, in the stress circle, of the principal plane. on which σ_2 acts, plane AD on the element.
$PM = $ the principal stress σ_1.
$OM = $ the position, in the stress circle, of the principal plane on which σ_1 acts, plane AC on the element.
$OR = $ the radius of the stress circle which, to scale, represents the maximum shear stress.
$\angle NOR = 2\theta = 2 \times \angle CAB$

All angles in the stress circle are double those on the element; ie the angle between the principal plane is 90° on the element, but the angle between OM and OL in the circle is 180°.

Although, as previously stated, problems may be solved by drawing the circle to scale to represent the various stresses, it is usually just as quick to sketch the diagram and calculate the values from the geometry of the figure. The examples which follow indicate a number of ways in which the circle can be used. In all cases the diagrams are drawn to scale but at the same time the various values are calculated. If only the scaled drawing is

90 Mechanical Science IV

used it is essential to use as large a scale as possible to ensure results of reasonable accuracy.

Example 4.10 (Only applied shear stress given)
At a point in a material there is a shear stress of 4 MN/m². Determine, using the Mohr circle method:
  (a) the principal stresses and the planes on which they act;
  (b) the normal, shear and resultant stress on a plane at 30° to the plane of the applied stress.

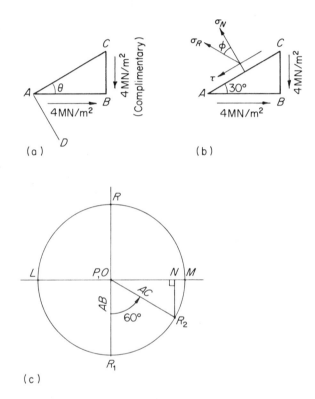

Fig. 4.23 (a) Principal stresses (b) Stress on an oblique plane (c) The Mohr stress circle

The shear stress on plane AB is negative and is represented by OR_1 in the stress circle.
The complimentary shear stress on plane BC is positive and is represented by OR in the stress circle.
The points P and O coincide as there are no normal stresses on AB or BC.
(a) $PM = OR = 4$ MN/m² *tensile* (i.e. to the right of P)
$PL = OR = 4$MN/m² *compressive* (i.e. to the left of P)
OR_1 is the plane of AB, OM the plane of PM which are 90° apart in the stress circle and thus 45° apart on the element.
So, in Fig. 4.23 (a) $\theta = 45°$, the stress on AC is 4 MN/m² tensile and the stress on AD is 4 MN/m² compressive, the principal stresses. These are results which are known to be accurate from Section 4.9 (iv).

(b) Referring to Fig. 4.23 (c). In the stress circle, the plane AC, 30° from plane AB on the element, is represented by radius OR_2 which is 60° from OR_1. Then:

$$\sigma_N = PN = OR \cos 30° = 4 \times 0.866 = 3.464 \text{ MN/m}^2$$
$$\tau = NR_2 = OR \sin 30° = 4 \times 0.5 = 2.0 \text{ MN/m}^2$$
$$\sigma_R = OR_2 = 4 \text{ MN/m}^2$$
$$\phi = \tan^{-1} \tau/\sigma_N = \tan^{-1} \frac{2.0}{3.464} = 30°$$

Example 4.11 (*Given 2 tensile principal stresses*)
A thin boiler shell is subject to a hoop stress of 20 MN/m² and a longitudinal stress of 10 MN/m². Determine:
  (a) the normal, shear and resultant stresses on a plane at 25° to the plane of the hoop stress;
  (b) the maximum shear stress and the planes on which it acts.

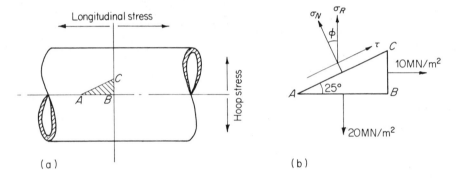

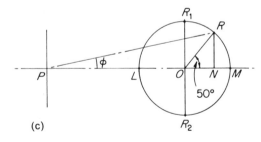

Fig. 4.24 (a) A boiler shell showing the stress planes (b) An enlarged element showing the principal applied stresses (c) The Mohr stress circle

(a) From Fig. 4.24(c)
$PL = 10 \text{ MN/m}^2$ and $PM = 20 \text{ MN/m}^2$ both principal stresses.
AB and BC are principal planes which are represented in the stress circle by OM and OL respectively.
Since plane AC is 25° anticlockwise from AB then in the stress circle OR, which represents plane AC, is 50° anticlockwise from OM, which represents AB.

$$LM = PM - PL = 20 - 10 = 10 \text{ MN/m}^2$$
$$OR = \tfrac{1}{2} LM = 5 \text{ MN/m}^2$$
$$ON = OR \cos 50° = 5 \times 0.643 = 3.125 \text{ MN/m}^2$$

(b) From Fig. 4.24(c)

Maximum shear stress = radius of stress circle = 5 MN/m²

The planes of maximum shear stress are at 45° to the principal planes (Section 4.10 ii) i.e. radii OR_1 and OR_2 in the stress circle.

Example 4.12 (Given one tensile and one compressive principal stress)

At a point in the wall of a pressure vessel there are principal stresses of 20 MN/m² tensile and 30 MN/m² compressive. Determine:
 (a) for a plane which is inclined at 50° to the plane of the smaller principal stress, the normal, shear and resultant stresses;
 (b) the maximum shear stress at the point.

Normal stress on plane $AC = PN$:

$$PN = PL + OL + ON = 10 + 5 + 3.22 = 18.22 \text{ MN/m}^2 \text{ tensile.}$$

Shear stress on plane $AC = RN$:

$$RN = OR \sin 50° = 5 \times 0.766 = 3.83 \text{ MN/m}^2$$

positive since R is above PM.

Resultant stress on plane $AC = PR$:

$$PR = \sqrt{18.22^2 + 3.83^2} = 18.6 \text{ MN/m}^2.$$

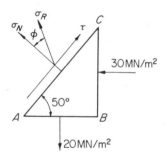

(a)

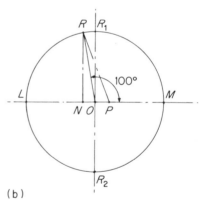

(b)

Fig. 4.25 (a) An element showing applied stresses (b) The stress circle

Referring to Fig. 4.25(b),

$$PM = 20 \text{ MN/m}^2 \text{ to scale,}$$

M is to the right of P since the stress is tensile.

$$PL = 30 \text{ MN/m}^2 \text{ to scale,}$$

L is to the left of P since the stress is compressive.

LM is then the diameter of the stress circle.

$\tfrac{1}{2}LM$ is the radius of the circle $= \dfrac{30 + 20}{2} = 25 \text{ MN/m}^2$

(a) In the circle, *OM* represents the plane *AB* on the element, so that *OR*, which is 100° anticlockwise from *OM* represents the plane *AC* which is 50° anticlockwise from *AB* on the element.

$$\therefore ROL = 80°$$

Shear stress on plane $AC = RN$:

$$RN = OR \sin 80° = 25 \times 0.985 = 24.6 \, MN/m^2$$

Normal stress on plane $AC = PN$, and $ON = 25 \cos 80° = 4.34 \, MN/m^2$:

$$PN = OP + ON$$
$$= (OM - PM) + ON$$
$$= (25 - 20) + 4.34 = 9.34 \, MN/m^2$$

Resultant stress on plane $AC = PR$:

$$PR = \sqrt{9.34^2 + 24.6^2} = 26.3 \, MN/m^2$$

(b) Maximum shear stress = radius of the stress circle = *OR* = 25 MN/m²

N.B. $\tau_{max} = \frac{1}{2}(\sigma_1 - \sigma_2)$ but since σ_2 is compressive it is written as $-30 \, MN/m^2$.

$$\therefore = \frac{1}{2}(20 - [-30]) = \frac{1}{2}(50) = 25 \, MN/m^2.$$

The planes of maximum shear stress are at 45° to the principal planes and in the stress circle are represented by OR_1 and OR_2.

Example 4.13 (*Deriving principal stresses from given compressive and shear stresses*)

A machine shaft 50 mm diameter is transmitting 30 kW at 1200 rev/min. It is also subject to an end thrust of 10 kN. Determine, for a point in the outer layers of the shaft,
  (a) the principal stresses and the inclination of the principal planes to the shaft axis;
  (b) the maximum shear stress.

$$\sigma_x = \text{compressive stress} = \frac{\text{end thrust}}{\text{cross-sectional area of shaft}}$$

$$= \frac{10 \times 10^3}{\pi/4 \times 0.05^2} = 5.1 \, MN/m^2$$

τ = shear stress due to torsion.

From the torsion formula $\dfrac{T}{J} = \dfrac{\tau}{r}$,

$$T = \frac{\text{power}}{\text{angular velocity (rad/s)}}$$

$$= \frac{30\,000}{1\,200 \times 2\pi/60} = 239 \, NM$$

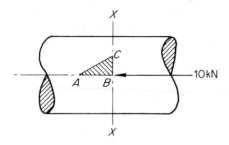

(a) Shaft showing stress planes

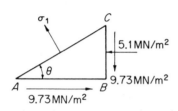

(b)

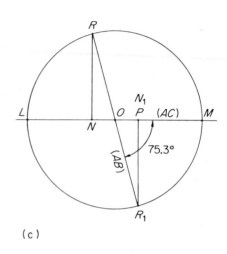

(c)

Fig. 4.26 (a) Shaft showing the stress planes (b) An element showing the applied stresses due to torsion and end thrust (c) The Mohr stress circle

$$J = \frac{\pi d^4}{32} = \frac{\pi \times 0.05^4}{32}$$
$$= 6.14 \times 10^{-7} \, \text{m}^4$$
$$r = 0.025 \, \text{m}.$$
$$\tau = \frac{Tr}{J}$$
$$= \frac{239 \times 0.025}{6.14 \times 10^{-7}}$$
$$= 9.73 \, \text{MN/m}^2$$

The stress circle (Fig. 4.26(c)) is then drawn.
$\sigma_x = 5.1 \, \text{MN/m}^2 = PN$ to scale with N to the left of P since σ_x is compressive.
$\tau = 9.73 \, \text{MN/m}^2 = NR$ to scale, R is above N since τ is positive.
$PR_1 = $ complimentary shear stress on plane $AB = 9.73 \, \text{MN/m}^2$, R_1 is below P.
There is no normal stress on plane AB so that points P and N_1 coincide. Joining R and R_1 bisects PN and gives the centre of the stress circle which has a radius of OR or OR_1.
(a) Principal stresses. These are obtained by scaling PM and PL or by simple calculation as follows.

$$ON = ON_1 = \tfrac{1}{2}PN = \frac{5.1}{2} = 2.55 \, \text{MN/m}^2$$

$$OM = OR = \sqrt{ON^2 + NR^2} = \sqrt{2.55^2 + 9.73^2}$$
$$= 10.1 \text{ MN/m}^2$$

Now $PM = OM - ON_1 = 10.1 - 2.55 = 7.55$ MN/m² *tensile*

$PL = OL + ON_1 = 10.1 + 2.55 = 12.65$ MN/m²

compressive.

Plane AB, parallel to the shaft axis, is represented in the stress circle by OR_1 since N_1R_1 is the shear stress on plane AB. OR_1 must be rotated anticlockwise through $\angle R_1OM$ to coincide with the plane of principal stress PM which is represented by OM.

$$\angle R_1OM = \tan^{-1}\frac{N_1R_1}{ON_1} = \tan^{-1}\frac{9.73}{2.55} = 75.3°.$$

∴ On the element, the plane AB must be rotated $\dfrac{75.3°}{2}$ to coincide with the plane of principal stress PM, i.e. plane AC.

Thus $\theta = 37.65°$ which is also the angle to the shaft axis.
The other principal plane is $37.65° + 90°$ or $127.65°$ to the shaft axis.

(b) The maximum shear stress is the radius of the stress circle to scale.

i.e. $\tau_{max} = 10.1$ MN/m²

The results are summarised in Fig. 4.27.

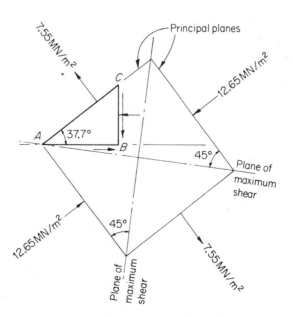

Fig. 4.27 A summary of the principal stresses and maximum shear stress

Example 4.14
At a point in a material there is a tensile stress of 20 MN/m² acting on a vertical plane together with a shear stress of 15 MN/m². On a horizontal plane there is a tensile stress of 40 MN/m² together with the complimentary shear stress. The stress in the third plane is zero. Determine:
 (a) the principal stresses and the planes on which they act;
 (b) the maximum shear stress.
Display the results on a suitable diagram.

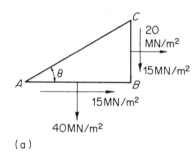

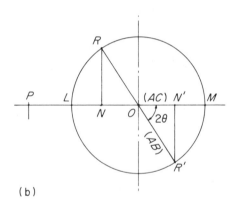

Fig. 4.28 (a) An element showing the applied stresses (b) The Mohr stress circle

Referring to Fig. 4.28(a) and (b):
$PN = 20$ MN/m² tensile; $NR = 15$ MN/m², positive shear stress on plane BC.
$\therefore OR$ represents plane BC in the stress circle.
$PN' = 40$ MN/m² tensile; $N'R' = 15$ MN/m², negative shear stress on plane AB.
$\therefore OR'$ represents plane AB in the stress circle.

$$ON = ON' = \tfrac{1}{2}(PN' - PN) = \tfrac{1}{2}(40 - 20) = 10 \text{ MN/m}^2.$$
$$OR = OL = OM = \sqrt{ON^2 + NR^2} = \sqrt{10^2 + 15^2}$$
$$= 18 \text{ MN/m}^2$$

(a) Principal stress $PM = PN + NO + OM = 20 + 10 + 18$
$\qquad\qquad\qquad\qquad = 48$ MN/m² *tensile*.
Principal stress $PL = PM - 2OM = 48 - 36$
$\qquad\qquad\qquad\qquad = 12$ MN/m² *tensile*.

The plane of the principal stress, PM, is OM so that $\angle MOR'$ in the stress circle is 2θ on the element.

Since $\tan \angle MOR' = \dfrac{N'R'}{ON'} = \dfrac{15}{10} = 1.5$ then $\angle MOR' = 56.3°$.

$$\therefore \theta = \frac{56.3}{2} = 28.15°.$$

The other principal plane is $28.15 + 90 = 118.5°$ from plane AC.

(b) The maximum shear stress is the circle radius to scale, i.e. 18 MN/m². This acts on planes at 45° to the principal planes. The results are displayed in Fig. 4.29.

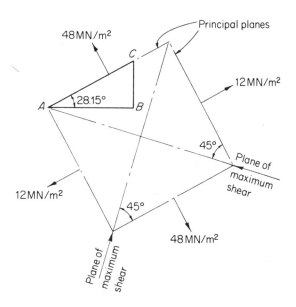

Fig. 4.29 A summary of the principal stresses, the planes on which they act and the maximum shear stress

From these examples it is apparent that provided the stress circle is drawn approximately to scale, the various angles and dimensions are easily calculated. An accurately scaled diagram will, at best, give results of very limited accuracy, but this may be acceptable in certain circumstances where approximations only are required.

4.12 The elastic constants

These were initially referred to in Section 4.7 when the concept of *bulk modulus* was introduced. Since we have now shown that tensile, compressive and shear stresses are very much interrelated it is understandable that the various elastic modulii should also be related. The two relationships which we shall now derive will enable approximations of the remaining constants to be obtained from only two elastic constants. The elastic constants are:

$$E = \text{modulus of elasticity or Youngs modulus}$$
$$= \frac{\text{tensile or compressive stress}}{\text{tensile or compressive strain}} = \frac{\sigma}{\varepsilon}$$

$$v = \text{Poisson's ratio} = \frac{\text{lateral strain}}{\text{longitudinal strain}}$$

$$G = \text{Modulus of rigidity} = \frac{\text{shear stress}}{\text{shear strain}} = \frac{\tau}{\phi}$$

$$K = \text{Bulk modulus} = \frac{\text{equal three dimensional stress}}{\text{volumetric strain}}$$

$$= \frac{-\sigma}{\theta}$$

(i) Relationship between E, *v* and K

For the cube of material shown in Fig. 4.30, which is subject to equal three dimensional compressive stress or hydrostatic pressure:

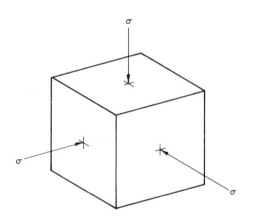

Fig. 4.30 A cube of material subject to hydrostatic pressure or equal stresses in three dimensions

volumetric strain = 3 × linear strain (Section 4.4)

$$\text{linear strain} = \frac{\sigma}{E} + \frac{v\sigma}{E} + \frac{v\sigma}{E} = -\frac{\sigma}{E}(1 - 2v)$$

$$\therefore \text{volumetric strain} = -\frac{3\sigma}{E}(1 - 2v) = \theta$$

The bulk modulus was defined in Section 4.7:

i.e. $K = \dfrac{-p \text{ or } -\sigma}{\theta}$

so, substituting for θ, the volumetric strain is

$$K = \frac{-\sigma}{\frac{3\sigma}{E}(1-2v)}$$

$$E = 3K(1-2v).$$

(ii) Relationship between E, G and v.

It must be remembered that the strains which are being considered are extremely small and that sketches, of necessity, magnify those strains very considerably. Thus the approximations which are made in deriving this relationship, are far more acceptable than Fig. 4.31(a) and (b) would suggest.

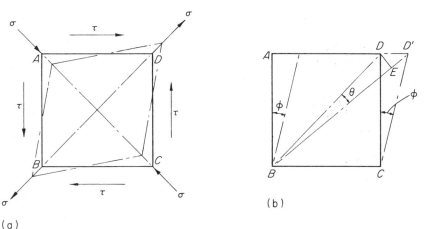

Fig. 4.31 Diagram to derive the relationship between E, G, and v

If an element *ABCD* is subject to a shear stress (τ) on faces *AD* and *BC* it has been shown that a complimentary shear stress (τ) will exist on faces *AB* and *DC* as shown in Fig. 4.31(a). From our consideration of complex stress systems, we also know that the shear stress results in principal stresses, one tensile and one compressive, of magnitude equal to the shear stress. These will act on planes at 45° to the shear planes, i.e. diagonally from corner to corner as shown once again in Fig. 4.31(a). Consequently, the diagonal *AC* is subject to a compressive strain and *BD* to tensile strain. The figure shows, exaggerated, the effect on the element.

For the purpose of this analysis and because the strains in practice are very small, it is assumed that the base *BC* remains unstrained and that *AD* moves relative to *BC*. The diagonal *BD* moves to *BD'* (Fig. 4.31(b)). From *D* a line is drawn perpendicular to *BD'* to cut it at *E*. For very small strains it is assumed that:

$$\angle D'DE = \angle ED'D = 45°.$$

Since $\angle DED' = 90°$,
$$DE = D'E = DD' \sin 45° = 0.707\, DD'.$$
Since DE is perpendicular to BD' so,
$$BD = BE.$$
The diagonal $BD = DC/\sin 45° = 1.414\, CD$.

Tensile strain on diagonal $BD = \dfrac{D'E}{BD} = \dfrac{0.707\, DD'}{1.414\, CD} = \dfrac{1}{2}\dfrac{DD'}{CD}$.

Shear strain on the element $= \dfrac{DD'}{CD} = \phi$ radians.

$\therefore$ Tensile strain $= \tfrac{1}{2}$ shear strain.

Taking into account both principal stresses:
$$\text{Tensile strain} = \frac{\sigma}{E} + \frac{v\sigma}{E} = \frac{\sigma}{E}(1+v)$$

$$\text{Shear strain} = \frac{\tau}{G}$$

$$\frac{2\sigma}{E}(1+v) = \frac{\tau}{G} \text{ but } \sigma = \tau$$

$$\therefore E = 2G(1+v)$$

Example 4.15
Tests on a new type of alloy suggested that its modulus of elasticity was $180\, GN/M^2$ and Poisson's ratio was 0.23.

(a) Determine, approximately, the modulus of rigidity and the bulk modulus.

(b) If a 50 mm cube of the material was subjected to a hydrostatic pressure of 200 bar, determine the change in volume.

(a) From $E = 3K(1-2v)$
$$K = \frac{E}{3(1-2v)} = \frac{180 \times 10^9}{3(1-0.46)} = 111 \times 10^9\, N/m^2$$
$$\text{or } 111\, GN/m^2$$

From $E = 2G(1+v)$
$$G = \frac{E}{2(1+v)} = \frac{180 \times 10^9}{2(1.23)} = 73.2 \times 10^9\, N/m^2 \text{ or}$$
$$73.2\, GN/m^2$$

(b) $$K = \frac{-p}{\theta} = \frac{-p}{\delta V/V}$$

$p = 200$ bar $= 20 \times 10^5\, N/m^2$
$\qquad = 2\, MN/m^2$
$v = 0.05\, m^3$

$$\therefore \delta V = \frac{pV}{K}$$

$$= \frac{2 \times 10^6 \times 0.05^3}{111 \times 10^9}$$

$$= 2.25 \times 10^{-9} \, m^3 \text{ or } 2.25 \, mm^3$$

4.13 Summary

All the stress theories are based on an *isotropic material*

(i) *Volumetric strain* $(\theta) = \delta V/V$.

(ii) *Poisson's ratio* $(v) = \dfrac{\text{lateral strain}}{\text{longitudinal strain}}$.

(iii) *Bulk modulus* $(K) = \dfrac{-\sigma \text{ or } -P}{\theta}$.

(iv) *Thin cylinder* — hoop stress $(\sigma_H) = \dfrac{pd}{2t}$;

— longitudinal stress $(\sigma_2) = \dfrac{pd}{4t}$;

— volumetric strain $= \dfrac{2}{E}(\sigma_H - v\sigma_2)$

$+ \dfrac{1}{E}(\sigma_L - v\sigma_H)$.

(v) *Thin spherical shells* — stress $= \dfrac{pd}{4t}$;

— volumetric strain $= \dfrac{3\sigma}{E}(1-v)$.

(vi) *Complimentary shear stress* = applied shear stress.
(vii) *Complimentary shear stress* acts on planes at 90° to the applied shear.
(viii) *Tensile and compressive stresses* result in shear stresses.
(ix) *Maximum shear stress* due to tension or compression = $\frac{1}{2}$ tensile or compressive stress.
(x) *Shear stresses* result in tensile and compressive stresses.
(xi) *Maximum tensile or compressive stress* due to shear stress is equal to that shear stress.
(xii) *Principal stresses* are the *maximum* or *minimum* normal stresses in a system, *i.e.* they are tensile and/or compressive.
(xiii) *Principal planes* are the planes on which the principal stresses act and they are planes of zero shear stress.
(xiv) *Planes of maximum shear stress* are at 45° to the principal planes.
$E = 3K(1-2v) = 2G(1+v)$

Exercises 4

Questions 1 to 5 refer to the theory and should be attempted initially without further reference to the text.

1. In connection with the elastic stress theories some of the following statements are accepted as true. Indicate the 'true' statements.
   (a) In practice a tensile force results solely in the elonggtion of the specimen.
   (b) A shear stress will result in a material being subject to both tensile and compressive stresses of equal magnitude.
   (c) A tensile stress also results in a material being subjected to a single shear stress of half its magnitude.
   (d) A shear stress on one plane will result in an equal but opposite shear stress on planes at right angles.
   (e) A material in compression will fail in shear if its shear strength is less than half its compressive strength.

2. Indicate which of the following statements are true.
   (a) An *isotropic* material has the same elastic properties at all points.
   (b) A *homogeneous* material has a uniform structure throughout.
   (c) An isotropic material need not be homogeneous.
   (d) An *orthotropic* material has different elastic properties in each of the three dimensions.
   (e) An *anisotropic* material has different elastic properties from point to point.

3. In connection with two-dimensional stress systems, indicate which of the following statements are correct.
   (a) A principal stress is either tensile or compressive.
   (b) A principal plane is a plane on which the shear stress is zero.
   (c) One principal stress is always tensile and the other is compressive.
   (d) The principal stresses are the maximum and minimum normal stresses in the system.
   (e) The maximum shear stress is always half the algebraic difference between the principal stresses.

4. In connection with the Mohr stress circle indicate which of the following statements are true.
   (a) Normal stresses are represented by horizontal lines and shear stresses by vertical lines.
   (b) The circle radius represents, to scale, the maximum shear stress.

(c) Compressive stresses are drawn to the left of the pole or starting point.
(d) Angles in the circle representing the stress planes are double those on the element.
(e) The circle diameter is half the difference between the principal stresses.

5 With regard to the elastic constants indicate the correct statements in the following.
(a) There are four elastic constants represented by the symbols: E, G, K and v
(b) E, K and v are related by the expression $E = 3K(1-2v)$
(c) E, G and v are related by the expression $E = 2G(1+2v)$
(d) The bulk modulus $K = \dfrac{\sigma}{\delta V/V}$
(e) The volumetric strain is the sum of the linear strains.

6 A tie bar is to carry a tensile load of 120 kN. Determine the diameter of the bar if:
(i) the tensile stress is not to exceed 50 MN/m²;
(ii) the shear stress is not to exceed 30 MN/m².

7 A short circular column is to be designed so that:
(i) the maximum compressive stress is limited to 80 MN/m²;
(ii) the maximum shear stress is limited to 50 MN/m².
Determine in each case the safe diameter for a load of 180 KN compressive.

8 At a point in a component there is a pure shear stress of 25 MN/m², determine:
(i) the normal, shear and resultant stress on a plane at 30° to that of the applied stress;
(ii) the maximum direct stresses and the planes on which they act.

9 A material is tested in compression until it fractures. The test piece is cylindrical and 20 mm diameter. Fracture occurs at a load of 25 kN. The fracture is a shear fracture along a plane at 35° to the vertical axis of the test piece. Determine the ultimate shear stress for the material

10 At a point in the shell of a pressure vessel there is a tensile stress of 45 MN/m² and a compressive stress of 35 MN/m² on planes at right angles. Determine:
(i) the normal and shear stresses on a plane at 35° to that of the tensile stress;
(ii) the resultant stress on the plane in (i) and its inclination to the normal;
(iii) the maximum shear stress.

continued

11 At a point in a component two tensile stresses of 30 and 70 MN/m² are applied on planes at right angles. Determine:
   (i) the normal, shear and resultant stresses on a plane making an angle of 40° to the plane of the larger stress;
   (ii) the maximum shear stress.

12 A 50 mm diameter shaft which is transmitting a torque is subjected to a torsion-stress of 15 MN/m² and at the same time to an end thrust of 53 kN. Determine:
   (i) the principal stresses set up;
   (ii) the planes on which the principal stresses act relative to the shaft axis;
   (iii) the maximum shear stress.

13 A component in a machine is subject to a compressive stress of 35 MN/m² on one plane and a tensile stress of 20 MN/m² on a plane at right angles. On the plane of the compressive stress there is also a shear stress of 25 MN/m² which acts in the positive sense. Determine:
   (i) the principal stresses;
   (ii) the principal planes relative to the plane of the compressive stress;
   (iii) the maximum shear stress.

14 At a point in a machine component there are tensile stresses of 3 MN/m² and 4 MN/m² together with shear stresses of 2 MN/m². Determine:
   (i) the principal stresses;
   (ii) the inclination of the principal planes to the plane of the 4 MN/m² stress;
   (iii) the maximum shear stress.

15 A component in a machine which is submerged in water is subjected as a result, to a hydrostatic pressure of 5 bar. The modulus of elasticity is 200 GN/m². If Poisson's ratio is 0.25 and the original volume was 20 cm³ determine:
   (i) the volumetric strain;
   (ii) the change in volume.

16 A component used in diving must operate at a depth of 60 m. It is made of a material for which the modulus of elasticity is 80 GN/m² and Poisson's ratio is 0.2. If the volume of the component is 0.25 m³ determine the change in volume at the full operating depth.
(Assume that the hydrostatic pressure = 10 000 × depth)

17 For both the components in questions in 15 and 16 determine the modulus of rigidity.

18 A support bar in a machine is 0.5 m long and 20 mm square cross-section. It is clamped along its length, between two

opposite faces, with a force of 50 kN and carries a longitudinal tensile load of 40 kN.
$E = 200 \text{ GN/m}^2$
$v = 0.33$
Determine:
    (i) the longitudinal strain;
    (ii) the lateral strain;
    (iii) the cross-sectional dimensions when the bar is strained.

19 A thin boiler shell is made of steel plate 20 mm thick. The nominal diameter is 3 m. For a steam pressure of 12 bar determine:
    (i) the stress induced on a longitudinal section;
    (ii) the stress induced on a circumferential section;
    (iii) the change in volume on a length of 5 m.
Assume that $E = 200 \text{ GN/m}^2$ and $v = 0.33$.

20 A spherical pressure vessel 3.5 m diameter is made of steel plate 10 mm thick. The maximum safe stress is 60 MN/m². Determine:
    (i) the maximum pressure to which the shell may be subjected;
    (ii) the change in volume at the maximum safe pressure.
$E = 200 \text{ GN/m}^2$ and $v = 0.3$.

5 Belt Drives

5.1 Introduction

In the design of machines the problems of transmitting power between parallel shafts often arises. The design may call for a speed modification between the shafts, it may require economy and quietness in operation and a measure of overload protection for the driving motor. In these circumstances, particularly if the shaft centre distance is large, some form of belt drive may be chosen.

Modern belt drive systems usually comprise a V-section belt running in a v-grooved pulley, although for a few applications both the flat belt and pulley and the round-section belt with a v-pulley still persist. Multigrooved v-pulleys provide additional power transmission with the freedom from complete breakdown should one belt break. Belt drives can absorb shock loads by stretching or slipping and thus provide some overload protection to driving motors.

Toothed belts are also in common use but they are not part of this theory as they do not rely on friction for the transmission of power. The theory is developed in the first instance for the flat belt.

5.2 Centripetal tension

Fig. 5.1 (a) depicts a belt stretched over two pulleys of equal diameter. If a drive is to be transmitted from one pulley to the other, then the portions of the belt *ABC* and *DEF* in contact with the pulleys must move in a circular path. The belt mass involved must, therefore, be subjected to a centripetal force in order to provide the necessary centripetal acceleration. This force can only be provided by tension in the belt material. For low belt velocities this tension is not significant and is often ignored. However with the high speed drives now in common use it cannot be disregarded.

Fig. 5.1 (b) shows a flat belt passing over a pulley which is rotating at ω rad/s. The mean radius of the belt in contact with the pulley is *r*. A small element of the belt subtending an angle $\delta\theta$ at

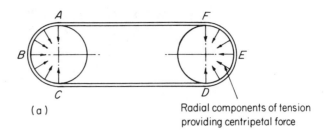

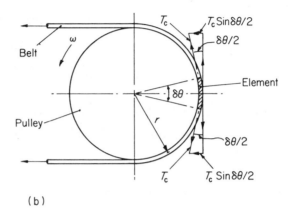

Fig. 5.1 (a) A belt stretched over two pulleys
(b) Centripetal tension in a flat belt passing over a pulley

the pulley centre is considered. The mass of unit length of belt material is m.

$$\text{length of the element} = r\,\delta\theta$$
$$\text{mass of the element} = mr\,\delta\theta$$
$$\text{velocity of the element} = w \text{ rad/s}$$
$$\text{centripetal acceleration} = \omega^2 r$$
$$\text{centripetal force} = \omega^2 r (mr\,\delta\theta)$$
$$= m\omega^2 r^2\,\delta\theta \text{ but since } v = \omega r$$
$$= mv^2\,\delta\theta. \tag{5.1}$$

The centripetal tension on the element is T_c. This is resolved into radial components each of which $= T_c \sin \dfrac{\delta\theta}{2}$. For small angles the sine approximately equals the angle in radians, so the radial components may be written as $T_c\,\delta\theta/2$.

Total radial or centripetal force provided
$$= 2T_c\,\delta\theta/2 \text{ or } T_c\,\delta\theta \tag{5.2}$$

Equating 5.1 and 5.2:
$$T_c\,\delta\theta = mv^2\,\delta\theta$$

i.e. $T_c = mv^2.$ (5.3)

Example 5.1
A flat belt is 10 cm wide and 0.4 cm thick and is to transmit power between two pulleys each of which is 0.3 m diameter. If the

pulleys are to run at 1 200 rev/min, determine the centrifugal tension in the belt. The belt material has a density of 0.001 kg/cm³

$$T_c = mv^2$$

m = mass/unit length
   = density × width × thickness × 1 m
   = 0.001 × 10 × 0.4 × 100 cm
   = 0.4 kg/100 cm or 0.4 kg/m

$$v = \omega r = 1200 \times \frac{2\pi}{60} \times 0.15$$
$$= 18.8 \text{ m/s}.$$
$$T_c = 0.4 \times 18.8^2$$
$$= 141 \text{ N}.$$

This is a significant force and represents a stress in the belt material of $\dfrac{141}{0.1 \times 0.004} = 352.5 \text{ kN/m}^2$.

5.3 Tension and power transmission

To transmit power, the belt must not slip relative to the pulleys. This means that in addition to the centrifugal tension, there must be some additional tension to provide a frictional grip between the belt and the pulley. This additional tension is achieved in practice by selecting a suitable belt length for a given centre distance between pulleys, and either providing some means of adjustment of centre distance or of otherwise tensioning the belt.

When the pulleys in Fig. 5.2 are stationary, the tension is the same on both sides T_0. As the driving pulley is then rotated, the belt will stretch and tighten between A and B due to resistance from the driven pulley. This continues until there is sufficient tension to overcome the resistance to motion. If the belt behaves elastically, i.e. it obeys Hooke's Law, there will be a corresponding slackening and contraction between C and D. Thus:
(i) there will now be two tensions, T_1 in the tight side and T_2 in the slack side;
(ii) the belt will remain the same overall length.

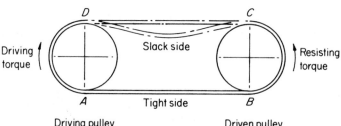

Fig. 5.2 The tensions in a belt on two pulleys

From (i):
$$\text{net force transmitted to driven pulley} = (T_1 - T_2)$$
$$\text{Power transmitted} = \text{force} \times \text{linear velocity}$$
i.e. $P = (T_1 - T_2) v$ (5.4)

or $P = (T_1 - T_2) \omega r$ (5.5)

where v is the linear velocity of the belt, or a point on the pulley rim, ω is the angular velocity of the pulley and r is its radius.

From (ii):
$$T_1 - T_0 = T_0 - T_2$$
$$\therefore (T_1 + T_2) = 2T_0 \quad (5.6)$$

N.B. In this part of the calculations T_c does not appear, providing it has been subtracted from the total permissible tension, before the calculations are commenced.

$$\text{Thus } T_1 = (\text{total safe tension} - T_c)$$

Example 5.2
In a belt drive the effective pulley diameter is 0·25 m. The belt section is 12 cm × 0.9 cm and the mass per unit length is 0.5 kg/m. If the belt velocity is 20 m/s; $T_2 = 0.4 T_1$ and the maximum stress in the belt material must not exceed 1 200 kN/m², determine:
 (a) T_1 and T_2;
 (b) the initial tension;
 (c) the power transmitted.

From Equation 5.3:
$$T_c = mv^2$$
$$= 0.5 \times 20^2$$
$$= 200 \text{ N}$$

Maximum safe belt tension = maximum stress × cross-sectional area of belt
$$= 1\,200 \times 10^3 \times 0.12 \times 0.009$$
$$= 1\,296 \text{ N}$$

$\therefore$ Maximum value of $T_1 = 1\,296 - T_c = 1\,296 - 200$
$$= 1\,096 \text{ N}$$
$\therefore T_2 = 0.4 \times 1\,096 = 438.4$ N

From Equation 5.6:
$$\text{Initial tension } T_0 = \frac{T_1 + T_2}{2}$$
$$= \frac{1\,096 + 438.4}{2} = 767.2 \text{ N}$$

The power transmitted is given by using Equation 5.4:
$$P = (T_1 - T_2) v$$
$$= (1\,096 - 438.4)\, 20$$
$$= 13\,152 \text{ W or } 13.152 \text{ kW}$$

5.4 Angle of lap, friction force, belt tensions

(i) Angle of lap (θ)

This is the angle subtended at the centre of the pulley by the arc over which the belt is in contact with the pulley. The length of arc will materially affect the frictional driving force which can be developed. Fig. 5.3 (a), (b) and (c) shows the possible arrangements and their effects on the angle of lap.

For pulleys of equal diameter, the angle of lap is 180° or π radians and is the same for both pulleys. For pulleys of unequal

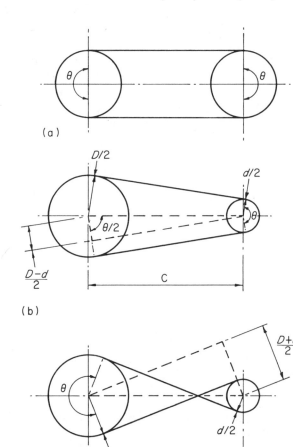

Fig. 5.3 (a) Equal pulleys
$\theta = 180°$ or π radians
(b) Unequal pulleys
$\theta = 2 \cos^{-1} \dfrac{D-d}{2C}$
(c) Crossed belt
$\theta = 360 - 2 \cos^{-1} \dfrac{D+d}{2C}$

diameter with the open belt arrangement of Fig. 5.3 (b), the angle of lap on the smaller pulley is the limiting factor, as it is smaller than the angle of lap on the larger pulley.

$$\text{Angle of lap } \theta = 2 \cos^{-1}\left(\frac{D-d}{2c}\right)$$

where D and d are the pulley diameters and c is their centre distance.

To increase the angle of lap in the case of flat or circular section belts (in v-grooved pulleys), the cross-belt arrangement of Fig. 5.3 (c) can be used. In this case the angle of lap is the same for both pulleys which, however, rotate in opposite directions.

$$\theta = 360 - 2 \cos^{-1}\left(\frac{D+d}{2c}\right)$$

Example 5.3
Two pulleys 0.5 m and 0.25 m diameter are connected by a flat belt. Calculate the angle of lap on the smaller pulley if the centre distance is 1.5 m and
   (a) the open-belt method is used;
   (b) the crossed belt method is used.

(a) For the open belt:

$$\theta = 2 \cos^{-1}\left(\frac{0.5 - 0.25}{2 \times 1.5}\right)$$
$$= 2 \cos^{-1} 0.083$$
$$= 2 \times 85.2° = 170.4° \text{ or } 2.97 \text{ radians*}$$

(b) For the crossed belt:

$$\theta = 360° - 2 \cos^{-1}\left(\frac{0.5 + 0.25}{2 \times 1.5}\right)$$
$$= 360° - 2 \cos^{-1} 0.25$$
$$= 360° - 151° = 209° \text{ or } 3.646 \text{ radians*}$$

*In the next section it will be seen that the angle of lap needs to be calculated in radians.

(ii) Friction force and belt tension

Fig. 5.4 (a) shows a pulley of radius r being driven by a flat belt in a clockwise direction. The blet tensions are T_1 and T_2 on the tight and slack sides respectively and the total angle of lap is θ radians. Considering the small element of the belt, which is shown subtending an angle $\delta\theta$ at the centre, let the tensile force (due to the belt tension) be T towards the slack side, and $(T+\delta T)$ towards the tight side. These forces will act at right angles to the end faces of the element and make an angle $\delta\theta/2$ with the tangent. They must both be resolved into tangential and radial com-

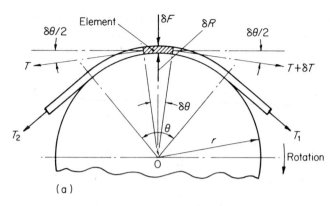

Fig. 5.4 (a) Tensions on the tight and slack sides of a belt drive
(b), (c) Vector diagrams resolving the tensions.

ponents. The vector diagrams Fig. 5.4 (b) and (c) show the respective components.

Radial components are: $T \sin \delta\theta/2$ and $(T+\delta T) \sin \delta\theta/2$.

Since for small angles sine = angle in radians (approximately) and small products such as $\delta\theta/2 \times \delta T$ can be ignored, the radial components can be written as $T \delta\theta/2$.

$$\therefore \text{Total radial force} = 2T\delta\theta/2$$
$$= T\delta\theta$$

The pulley reaction δR arises from this force so that:

$$\delta R = T \delta\theta.$$

If μ is the coefficient of friction between the belt and the pulley then:
$$\text{Tangential friction force} = \mu \delta R = \mu T \delta\theta$$

This is the force which provides the power transmission and which, if no slip occurs, must equal the net tangential force acting on the element.

From Fig. 5.4 (b) and (c) once again:

Net tangential force = $(T+\delta T)\cos \delta\theta/2 - T \cos \delta\theta/2$

Since $\dfrac{\delta\theta}{2}$ is very small it may be assumed that $\cos \dfrac{\delta\theta}{2} = 1$.

$$\therefore \text{Net tangential force} = (T+\delta T) - T = \delta T$$

Equating the net tangential force and the tangential friction force:

$$\delta T = \mu T \delta\theta \text{ or } \frac{\delta T}{T} = \mu \delta\theta$$

Over the whole angle of lap, θ, T varies from T_1 to T_2 so that by integrating both sides:

$$\int_{T_1}^{T_2} \frac{\delta T}{T} = \int_0^\theta \mu \delta\theta$$

$$\log_e \frac{T_1}{T_2} = \mu\theta \text{ or } \frac{T_1}{T_2} = e^{\mu\theta} \qquad (5.7)$$

This is a very important relationship between the angle of lap, in radians, and the belt tensions.

Example 5.4
A belt drive must transmit 7.5 kW at a belt speed of 12 m/s. The tension in the tight side of the drive must not exceed $2\frac{1}{4}$ times that in the slack side and 14 N/mm of belt width. Determine:
(a) the belt width;
(b) the angle of lap if $\mu = 0.3$

(a) From Equation 5.4:
$$P = (T_1 - T_2) v$$
$$T_2 = T_1/2.25 = 0.444\, T_1$$
$$v = 12 \text{ m/s}$$
$$P = 7\,500 \text{ W}$$
$$\therefore 7\,500 = (T_1 - 0.444\, T_1)\, 12$$
$$T_1 = \frac{7\,500}{12 \times 0.555}$$
$$= 1\,125 \text{ N}$$

$$\text{Belt width} = \frac{T_1}{14} = \frac{1\,125}{14} = 80.4 \text{ mm}$$

(b) From Equation 5.7:
$$\frac{T_1}{T_2} = e^{\mu\theta}$$
$$T_1/T_2 = 2.25$$
$$\mu = 0.3$$
$$\therefore \log_e \frac{T_1}{T_2} = \mu\theta$$
$$\theta = \frac{\log_e 2.25}{0.3} = 2.7 \text{ radians or } 155°.$$

5.5 Maximum power transmission

In Section 5.3 it was established by Equation 5.4 that:
$$p = (T_1 - T_2) v.$$

It has just been established by Equation 5.7 that:
$$\frac{T_1}{T_2} = e^{\mu\theta}.$$
From these two equations it is now possible, for a given belt drive system, to determine the maximum power which can be transmitted. For this purpose it is necessary to recall Equation 5.3 for centrifugal tension:
$$T_c = mv^2$$

From
$$\frac{T_1}{T_2} = e^{\mu\theta}$$
$$T_2 = T_1/e^{\mu\theta}$$

This value is substituted in the power formula, 5.4.
$$\therefore P = (T_1 - T_1/e^{\mu\theta})v$$
$$= T_1(1 - 1/e^{\mu\theta})v$$
$$= kT_1 v \text{ where } k = (1 - 1/e^{\mu\theta}).$$

Now $T_{max} = T_1 + T_c$ so that $T_1 = T_{max} - T_c$.
$$\therefore P = k(T_{max} - T_c)v.$$

As $T_c = mv^2$,
$$P = kvT_{max} - km v^3.$$

Differentiate and equate to zero to find the maximum value:
$$dP/dt = kT_{max} - 3kv^2 m = 0$$
$$k(T_{max} - 3mv^2) = 0$$
$$T_{max} = 3mv^2$$
$$\therefore T_{max} = 3T_c \qquad (5.8)$$

The relationship should only be used as a general guide to the maximum power which in practice is seldom achieved due to the need for safety margins etc.

Example 5.5

A flat-belt drive has unequal diameter pulleys so that the angle of lap is 4.84 radians on one pulley and 2.1 radians on the other. The mass of the belt material is 0.7 kg/m length and the coefficient of friction is 0.19. The breaking strength of the belt is 840 N. Determine for maximum power transmission:
 (a) the linear velocity of the belt;
 (b) the power transmitted.

From Equation 5.8:
$$T_{max} = 3T_c$$
$$= 3mv^2$$
$$\therefore v^2 = \frac{T_{max}}{3m} = \frac{840}{3 \times 0.7}$$
$$\therefore v = \sqrt{400} = 20 \text{ m/s}$$

which is the linear velocity of the belt.

$$T_c = \frac{T_{max}}{3} = \frac{840}{3} = 280 \text{ N}$$

$$\therefore T_1 = (T_{max} - T_c)$$
$$= 840 - 280 = 560 \text{ N}.$$

From Equation 5.7:

$$\frac{T_1}{T_2} = e^{\mu\theta}$$

$\theta = 2.1$ rad — the smaller angle of lap.
$\mu = 0.19$.

$$\therefore T_2 = \frac{560}{e^{0.19 \times 2.1}}$$
$$= 376 \text{ N}.$$

From Equation 5.4:

$$P = (T_1 - T_2) v$$
$$= (560 - 376) 20$$
$$= 3680 \, w \text{ or } 3.68 \, kw$$

Once again it would not be practical to achieve this output as no factor of safety has been allowed and it is quite likely that the belt would break before this speed was achieved.

5.6 V-pulleys and V-belts

The v-grooved pulley was originally used with a round-section belt made of cotton or other textile materials. It is generally used now with an endless v-section belt made of rubber impregnated materials which give greater reliability and a more positive drive. The theory developed for the round section belt is often applied to the v-belt. However in any particular design situation more empirical formulae are used which allow margins of safety for the strains set up due to bending the belt into a small arc etc. These are the subject of various British Standard Specifications, for example BSS 1440 1971, 3790 1973. It should be remembered that the theory assumes that the belts are only in contact with the sides of the groove and not the bottom.

Referring to Fig. 5.5 (a) and (b), which show, respectively, a round and a v-belt in a v-grooved pulley:

N is the normal reaction where the round belt contacts the pulley or the approximate normal resultant reaction between the v-belt and he pulley sides.

R is the radial reaction which is derived from the normal forces.

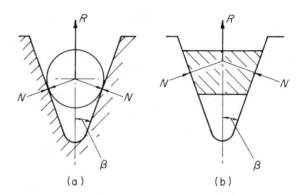

Fig. 5.5 (a) Round belt (b) V-belt

μ is the coefficient of friction between the belt and pulley.
β is the semi-groove angle of the vee.
The frictional force which causes the pulley to rotate is made up of two equal forces derived from the normal reactions N.

i.e. Friction force = $2N\mu$

From the relationship between N and R shown in Fig. 5.5 (a):

$$R = 2N \sin \beta \text{ so that } 2N = \frac{R}{\sin \beta} \text{ or } R \operatorname{cosec} \beta$$

$\therefore$ Friction force = $2N\mu = \mu R \operatorname{cosec} \beta$

In Section 5.4 if μR is replaced by $\mu R \operatorname{cosec} \beta$ as the tangential friction force then:

$$\frac{T_1}{T_2} = e^{\mu \theta \operatorname{cosec} \beta} \qquad (5.9)$$

Since $\operatorname{cosec} \beta$ is always greater than 1, except when $\beta = 90°$ and $2\beta = 180°$ i.e. a flat pulley, the friction effect is magnified by the vee. The term $\mu \operatorname{cosec} \beta$ is sometimes referred to as the *virtual coefficient of friction* (μ_1) for the v-pulley.

All the other formulae derived for the flat-belt drive apply to the v-pulley drives. In many cases multigroove pulleys are used with multiple belts in order to increase the power transmission. In such cases calculations are based on one belt and the result is multiplied according to the number of belts.

Most v-belts have an included angle of 40° but the pulley groove angle varies with the diameter of the pulley to allow for the fact that the belt will bulge more as it is bent round a pulley of smaller diameter. Groove angles of 32°, 34°, 36° and 38° are commonly specified.

Example 5.6
A rope pulley has a mean diameter of 0.5 m measured to the centre of the rope each side. The angle of lap is 190° and the included groove angle is 45°. The pulley rotates at 90 rev/min. The safe tension in one rope is 600 N and the coefficient of friction between the rope and the pulley is 0·25. Determine:

(a) the torque transmitted per rope;
(b) the number of ropes needed to transmit 2·5 kW.
Centrifugal effects may be ignored.

From $v = \omega r$:
$$\omega = 90 \times 2\pi/60$$
$$= 3\pi \text{ rad/s}$$
$$r = 0.5/2 = 0.25 \text{ m}$$
$$\therefore v = 3\pi \times 0.25$$
$$= 2.36 \text{ m/s}$$

$\theta = 190°$ or 3.32 rad.; $\beta = \dfrac{45}{2} = 22.5°$; cosec $22.5° = 2.61$.

$$\therefore \mu_1 \text{ or } \mu \operatorname{cosec} \beta = 0.25 \times 2.61 = 0.653$$
$$T_1 = 600 \text{ N ignoring centripetal effects.}$$

From Equation 5.9:
$$\frac{T_1}{T_2} = e^{\mu\theta \operatorname{cosec} \beta}$$

$$T_2 = \frac{T_1}{e^{\mu\theta \operatorname{cosec} \beta}} = \frac{600}{e^{0.653 \times 3.32}} = 68.6 \text{ N}$$

Torque $= (T_1 - T_2) \times$ radius $= (600 - 68.6) \, 0.25$
$$= 133 \text{ Nm per rope}$$

Torque to transmit 2.5 kw at 3π rad/s:
$$= \frac{P}{\omega} = \frac{2500}{3\pi} = 265.3 \text{ Nm}$$

$$\therefore \text{ No. of ropes} = \frac{265.3}{133} \text{ i.e. 2 ropes.}$$

Example 5.7
A belt drive has v-belts running in grooves of included angle 38°. Each belt has a working tensile strength of 200 N and a mass of 0.1 kg/m of length. The effective diameters of the two pulleys are 0.16 m and 0.48 m with a centre distance of 0.5 m. The belt speed is 12 m/s when the drive transmits 7 kW. The coefficient of friction between the belt and pulley is 0.15. Determine:
(a) the effective angle of lap;
(b) the number of belts required.

Referring Fig. 5.6:
$$\cos \theta/2 = \frac{0.16}{0.5} = 0.32$$
$$\therefore \theta/2 = 71.33° \text{ or } 1.245 \text{ rad}$$
$$\theta = 142.7° \text{ or } 2.49 \text{ rad on the smaller pulley}$$

118 Mechanical Science IV

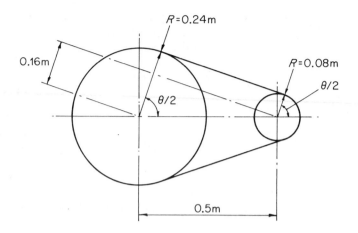

Fig. 5.6 Diagram of the V-belt drive

$$T_{max} = 200 \text{ N}$$
$$T_c = mv^2 = 0.1 \times 12^2 = 14.4 \text{ N.}$$
$$\therefore T_1 = (T_{max} - T_c) = (200 - 14.4) = 185.6 \text{ N}$$

Now $\beta = \dfrac{38}{2} = 19°$; cosec $\beta = 3.07$; $\mu = 0.15$

$$\therefore \mu\theta \text{ cosec } \beta = 0.15 \times 2.49 \times 3.07 = 1.147.$$

From Equation 5.9:

$$\frac{T_1}{T_2} = e^{\mu\theta \text{ cosec } \beta}$$

$$T_2 = \frac{T_1}{e^{\mu\theta \text{ cosec } \beta}} = \frac{185.6}{e^{1.147}}$$

$$\therefore T_2 = 59 \text{ N}$$

From Equation 5.4:

$$P = (T_1 - T_2) v$$
$$= (185.6 - 59) 12 = 1520 \text{ W/belt.}$$

$\therefore$ No of belts to transmit 7 kW

$$= \frac{7\,000}{1\,520} = 4.6 \text{ i.e. 5 belts.}$$

Example 5.8

Using the data given in Example 5.7 determine the maximum power which could be developed by the five belts and the velocity at which this would occur, if the initial tension can be adjusted.

$$T_{max} = 3 \, mv^2 = 0.3 \, v^2 = 200 \text{ N.}$$

$$\therefore v = \sqrt{\frac{200}{0.3}} = 25.8 \text{ m/s}$$

which is the velocity for maximum power.

$$\therefore T_c = mv^2 = 0.1 \times 25.8^2 = 66.7 \text{ N.}$$
$$T_1 = (T_{max} - T_c) = 200 - 66.7 = 133.3 \text{ N}$$
$$T_2 = \frac{T_1}{e^{1.147}} = \frac{133.3}{3.148} = 42.3 \text{ N}$$
$$\therefore P_{max} = (T_1 - T_2) v = (133.3 - 42.3) \, 25.8$$
$$= 2{,}348 \text{ W/belt}$$

For 5 belts the maximum power = 11 740 W or 11.74 kW.

5.7 Summary

(i) *Centrifugal tension* = mv^2.
(ii) *Power transmitted* = $(T_1 - T_2) v$ or $(T_1 - T_2) \omega r$.
(iii) *Tension ratios*: $T_1/T_2 = e^{\mu\theta}$ for a flat pulley.
$T_1/T_2 = e^{\mu\theta \operatorname{cosec} \beta}$ for a v-pulley.
(iv) *Maximum power* when $T_{max} = 3T_c$.

Exercises 5

Questions 1 to 5 refer to the theory and should be attempted initially without further reference to the text.

1 Belt drives have certain advantages over other forms of drive in a number of the following circumstances. Indicate which are correct.
   (a) They are quiet and efficient.
   (b) They will tolerate shock or excess loads.
   (c) They provide a speed modification.
   (d) The drive is positive.
   (e) They are expensive.

2 Multigrooved V-pulleys are often used in belt drives, in this connection indicate which of the following statements are correct.
   (a) They can be used with both 'vee' and round-section belts.
   (b) They provide quieter power transmission.
   (c) They guard against complete breakdown.
   (d) The drive is more positive than a flat belt.
   (e) The same basic theory as for flat belts is used in their design.

continued

3 Indicate which of the following relationships with regard to belt tensions are correct.
   (a) Maximum safe tension − centrifugal tension = tension in the tight side.
   (b) Centrifugal tension = mass/unit length of belt × velocity.
   (c) Power transmitted = (tight side tension − slack side tension) × belt velocity.
   (d) Tight side tension/slack side tension = $e^{\mu\theta}$.
   (e) Maximum power transmission when $T_{max} = 3$ × centrifugal tension.

4 The angle of lap in a belt drive:
   (a) is the angle, in radians, over which the belt is in contact with the pulley.
   (b) is the same for equal diameter driving and driven pulleys.
   (c) is the same for both pulleys when the belts are crossed.
   (d) limits the power transmission.
   (e) varies with the pulley centre distance.
   Indicate the correct statements.

5 The maximum power transmission for a given belt drive is:
   (a) limited by the strength of the belt material.
   (b) seldom achieved in practice.
   (c) depends upon belt velocity squared.
   (d) depends upon the angle of lap.
   (e) depends upon the density of the belt material.
   Indicate the correct statements.

6 A belt which is 7.5 cm wide and 0.3 cm thick is made of material which has a density of 0.001 kg/cm³. It transmits a drive between two pulleys of equal 0.25 m diameter. The pulleys run at 2 400 rev/min. If the maximum safe tension in the belt is 570 N determine:
   (i) the centrifugal tension;
   (ii) the tension on the tight side.

7 A flat-belt drive has an effective pulley diameter of 0.2 m and uses a belt whose cross-section is 10 cm × 0.75 cm with a mass of 0.5 kg/m. The belt velocity is 16 m/s and T_2 is to be half T_1. The stress in the belt material must not exceed 1 MN/m², determine:
   (i) T_1 and T_2;
   (ii) the initial tension;
   (iii) the power transmitted.

8 A flat-belt drive is to transmit 7.5 kW with a maximum belt velocity of 20 m/s. The stress in the belt material must not exceed 1.2 MN/m². The belt material has a density of 0.0012 kg/cm³. If $T_2 = 0.4\, T_1$ determine:

(i) the necessary width of belt 8 mm thick;
(ii) the centrifugal tension.

9 Two pulleys 0.6 m and 0.3 m diameter are connected by a flat belt. The coefficient of friction is 0.3 and T_1 must not exceed 800 N. Ignoring centrifugal tension, determine the power transmitted when the smaller, driving, pulley is running at 800 rev/min. The pulley centre distance is 1.2 m.

10 For the belt drive described in question 9 determine the power if the belts were crossed.

11 A flat-belt drive is to be designed to transmit 10 kW with a belt velocity of 15 m/s. The tension in the tight side must not exceed three times that in the slack side and the maximum tension must not be greater than 15 N/mm of belt width. The belt material has a density of 0.001 kg/cm^3. Determine:
   (i) the belt width;
   (ii) the angle of lap if $\mu = 0.35$;
   (iii) the maximum stress when transmitting 10 kW, if the belt is 10 mm thick.

12 A V-belt drive has belts running in grooves of 40° included angle. Each belt has a working tensile strength of 250 N and a mass of 0.1 kg/m. The effective pulley diameters are 180 mm and 500 mm with a centre distance of 600 mm. If the belt velocity is 15 m/s when 8 kW is being transmitted and $\mu = 0.15$ determine:
   (i) the angle of lap;
   (ii) the number of belts required.

13 A machine tool is driven by an electric motor by means of a v-belt. The pulley on the motor shaft has a diameter of 0.15 m whilst that on the machine tool gearbox shaft is 0.2 m diameter. The included V-angle is 40°. The motor runs at 1 500 rev/min. The maximum permitted belt tension of 1 kN gives a maximum power of 9 kW, determine:
   (i) the angle of lap for a pulley centre distance of 0.5 m;
   (ii) the tension on the slack side;
   (iii) the coefficient of friction.

14 A rope drive is to be designed to transmit a maximum power of 60 kW from a shaft running at 850 rev/min, by means of a number of ropes. The maximum tension in the ropes is to be 1 kN and their mass/unit length 0.75 kg/m. The angle of lap is 170° and the pulley grooves have an included angle of 40°.
If the coefficient of friction is 0.25 determine:
   (i) the effective pulley diameter;
   (ii) the rope tensions and the number of ropes to transmit the 60 kW.

continued

15 A compressor is driven by an electric motor by means of 2 v-belts. The belts each have a cross-sectional area of 3 cm², and the material of which they are made has a density of 0.0015 kg/cm³. The maximum stress in the belts is not to exceed 2.5 MN/m². The motor pulley has a diameter of 0.1 m and the compressor shaft pulley a diameter of 0.15 m. The pulley grooves have an included angle of 38° and the pulley centre distance is 0.4 m. If the coefficient of friction is 0.25 determine:
    (i) the maximum motor pulley speed for maximum power;
    (ii) the maximum power;
    (iii) the belt tensions for maximum power.

6 Mechanisms and Velocity Diagrams

6.1 Introduction

In the design of machine components it is necessary to know the forces which they may be required to transmit. These forces may be not only tensile, compressive or shear but at the same time cyclical, suddenly applied or gradually varying. Their effects can be quite different.

The piston, connecting rod and crank mechanism used in conjuction with most automobile engines, is a good example of a mechanism subject to varying forces. Considering the connecting rod alone, as one element, this transmits to the crankshaft the forces developed in the cylinder during the power or expansion stroke. These will vary from zero to a maximum. During the remainder of the cycle it will transmit varying forces from the crankshaft to the piston to enable the processes of exhaust, induction and compression to take place. At certain instants the connecting rod will be stationary, at others it will have a maximum velocity. As the rod possesses mass, it will itself be subject to accelerating and decelerating forces in addition to the forces it transmits. For quite a simple component this is a complicated design problem. In this chapter we shall be concerned with the first stage of such design, which is to determine the variation in velocity of the components or links in a mechanism. In later work it will then be possible to determine the accelerations and the forces and stresses which result.

6.2 Relative velocity

In a sense, all velocities are *relative* in that they are determined with respect to some original reference point. Thus an observer stationary on the ground will have a different impression of the velocity of an aircraft passing overhead from an observer in another aircraft travelling in the opposite direction, or even in the same direction. Similarly, in a mechanism velocities are determined relative to both stationary points and moving ones. They are indeed all essentially relative velocities.

6.3 Links, pairs, chains and mechanisms

A mechanism is constructed from a number of elements, generally called *links*, which are joined together in different ways to form *pairs*. Such pairs may take a number of forms, some of which are illustrated in Fig. 6.1 (a) to (c). The significance of the different types of pair is beyond the scope of this text but does become important in later work.

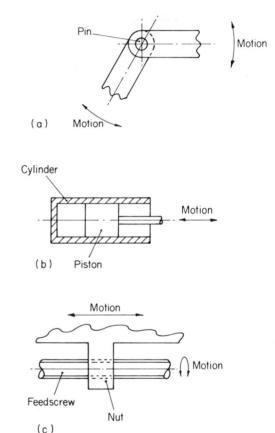

Fig. 6.1 (a) Turning pair (b) Sliding pair (c) Screwed pair

A number of pairs joined together form a *chain* and it is from the chain that a mechanism is derived. A mechanism is in fact a chain in which one link is fixed. This results in the other links being forced or constrained to move in a particular way. What, then, is the simplest mechanism which can be formed from a chain? Fig. 6.2 shows three links joined to form a chain, each link is part of two of the three pairs which are formed.

Only a very brief consideration is required to realise that such a chain is useless for transmitting motion since a rigid structure has been formed and the fixing of any one link will fix the whole structure. It is necessary therefore to add a fourth link to the chain

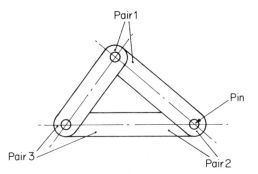

Fig. 6.2 3-bar chain

to produce the basis of the simplest of the mechanisms. This is called the *quadric cycle* or *four bar chain*.

Referring to Fig. 6.3, if the link *AD* is fixed and the link *AB* is rotated about the pin at *A* then the links *BC* and *CD* must move in a particular manner and their velocities will vary from instant to instant depending upon the position of *AB*. On the other hand if link *AB* is fixed and *AD* is retated about *A*, although *BC* and *CD* must move in a particular manner, with a varying velocity, their motion will be quite different from the previous case. Thus by fixing each link in turn and rotating one of the remaining links, the chain can be given a number of resultant motions. Each different arrangement is called an *inversion* of the basic chain. Other inversions may be produced by using sliding pairs instead of turning pairs. The crank and connecting rod arrangement referred to in Section 6.1 is in fact an inversion of the four bar chain known as the *slider crank chain*.

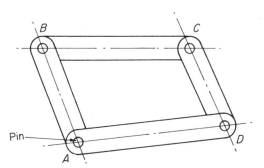

Fig. 6.3 4-bar quadric cycle chain

So much for the basic idea of a mechanism, it is now necessary to examine in more detail the actual relative velocities of the different parts of the mechanism.

6.4 Relative motion of a link

In Fig. 6.4 *AB* is a rigid link moving in the plane of the paper. The instantaneous velocity of end *B* is v_b in the horizontal direction,

Fig. 6.4 Diagram of the link AB and fixed point 'O'

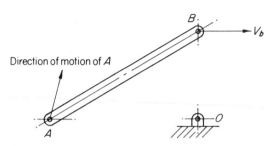

whilst the instantaneous motion of end A is in the direction shown but the velocity is unknown.

Choosing a suitable scale, in the case of a numerical example, we may represent the velocity of B, relative to the fixed point O, by the vector **ob**. (i.e. an observer at O would describe the velocity of B as being in the direction o to b and of magnitude, to scale, equal to the length of **ob**. On the other hand, if the observer was moving with B then the point O would appear to be moving with a velocity of the same magnitude but in the direction b to o. Thus the vector **ob** represents the instantaneous velocity of B relative to O. Thus the vector **bo** represents the instantaneous velocity of O relative to B. If this is understood no arrows are necessary. (Fig. 6.5 (a).)

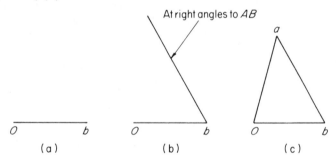

Fig. 6.5 Vector diagrams to determine the movement of the ends of the link relative to the fixed point

Rejoining the observer at B and looking at the end A of the solid link, then, relative to B, there can be only one motion. Since A cannot move directly towards or away from B, they are joined by the solid link, it can only move at right angles to the straight line joining them. This velocity can be represented by a vector from b in a direction at right angles to AB. For the moment its length is unknown. (Fig. 6.5 (b).)

Now once again, let us revert to the position O and observe the movement of A. Its direction is known so that a vector, originating at o, drawn in this direction will intersect the vector from b at what must be the point a. The velocity of A relative to B is now defined by the scaled length of vector **ba**. The velocity of A relative to O is defined by the scaled length of vector **oa**. This can be summarized as follows.

> The vector triangle *oab* enables us to define, in both magnitude and direction, the instantaneous motion of each point O, A and B relative to each other (Fig. 6.5(c)).

Magnitude = length to scale of the vector.
Direction is from the point of reference (i.e. from where the motion is observed) to the point being observed.

Thus *B* relative to *A*, direction *a* to *b*.
Thus *A* relative to *B*, direction *b* to *a*.

If this is clearly understood then the subsequent work will present little difficulty.
Note that capital letters are used to define points on the mechanism and the corresponding lower-case letters on the vector diagram represent their velocities.

Example 6.1
The connecting rod in an engine is, at a certain instant, inclined at 8° from the vertical, (Fig. 6.6 (a)). The gudgeon-pin end (*B*),

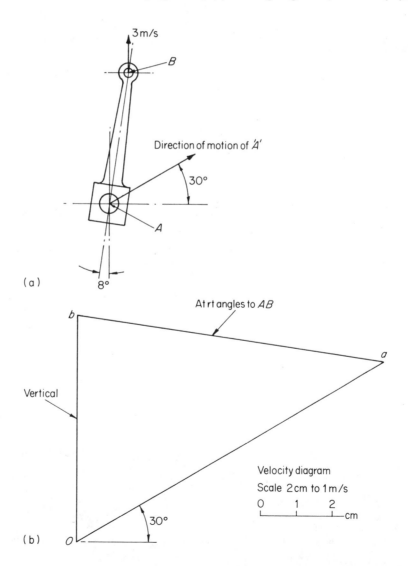

Fig. 6.6 (a) Space diagram
(b) Velocity diagram

attached to the piston, is moving vertically upwards at 3 m/s whilst the crankshaft end, defined by the point A, is at the same instant moving 30° above the horizontal to the right. Determine:
(a) the magnitude and the direction of the velocity of A relative to B.
(b) the magnitude of the velocity of A.

Note on scales. The scales quoted are those recommended to give reasonable accuracy for a graphical solution. For printing purposes the diagrams are not reproduced full size but the linear scale indicates the printed equivalent of 1 cm.

ob is drawn vertical, 6 cm long, to represent the velocity of B relative to the fixed point O. The point b is above o since B is moving upwards. Through o a line 30° above the horizontal is drawn to intersect a line through b at right angles to the line joining A and B. The point of intersection is a. Scaling **ab** and **oa** gives:
(a) velocity of A relative to B = 4.22 m/s in a direction from b to a;
(b) velocity of A relative to O = 4.82 m/s.

6.5 Angular velocity of a link

Velocity diagrams represent the instantaneous *linear* velocities of one point in a mechanism relative to any other point. Within a mechanism a link is pinned at its end(s) so that there will be an instantaneous rotation about the pins.

$$\text{Angular velocity} = \frac{\text{instantaneous relative linear velocity}}{\text{distance between pins or points}} \text{ rad/s}$$

Referring to Fig. 6.7 (a) and (b) showing a link AB:

linear velocity of B relative to A = v_{ab}

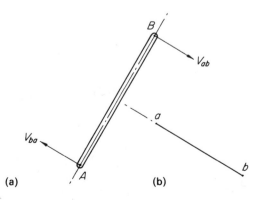

Fig. 6.7 (a) Space diagram
(b) Velocity diagram

which is the scaled length of the vector **ab**.

$$\therefore \text{Angular velocity of link} = \frac{v_{ab}}{AB} \text{ rad/s}$$

This is obviously a clockwise rotation about A. Similarly if the angular velocitiy is given, the linear velocity is obtained by reversing the procedure:

$$\text{i.e. } v_{ab} = \omega \times AB$$

6.6 Rubbing velocity at pins

As explained in the previous section, links are often joined together by pins. This results in a friction surface where the pin, fixed to one link, bears on the whole surface machined in the other link. The relative velocities of these two surfaces will influence the wear which takes place and the frictional resistance set up. The term 'pin' is a general one and in the case of the engine connecting-rod mentioned previously, it would mean the gudgeon-pin and the crankshaft journal.

To find the rubbing velocity:
 (i) determine the angular velocities of both links and their directions;
 (ii) determine the pin radius.

Then referring to Fig. 6.8 (a) to (c) since $v = \omega r$:

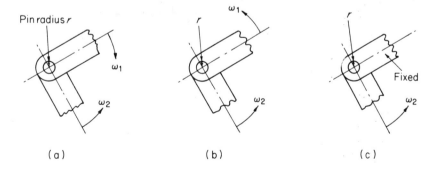

Fig. 6.8 Rubbing velocities (a) (b) (c)

in case (a): resultant relative angular velocity =
$= (\omega_1 + \omega_2)$ rad/s
rubbing velocity $= (\omega_1 + \omega_2)r$
in case (b): resultant relative angular velocity
$= (\omega_1 - \omega_2)$ rad/s
rubbing velocity $= (\omega_1 - \omega_2)r$
in case (c): resultant relative angular velocity $= (0 + \omega)$ rad/s
rubbing velocity $= (0 + \omega)r$

(If r is in metres then the rubbing velocity will be in m/s.)

Example 6.2
Referring to Example 6.1 if the gudgeon-pin was 1.5 cm diameter, the crankshaft journal 6 cm diameter and the length *AB* was 30 cm, determine the rubbing velocities in each case.
It may be assumed that the crankshaft journal has a clockwise velocity of 60 rad/s.

Linear velocity of connecting rod = 4.22 m/s = $\frac{4.22}{0.3}$

= 14.1 rad/s.

Direction at the gudgeon-pin end is anticlockwise.
Direction at the crankshaft end is clockwise.

∴ Rubbing velocity at the gudgeon-pin = (0 + 14.1) 0.0075
  = 0.106 m/s (case c).
∴ Rubbing velocity at the crankshaft = (14.1 + 60) 0.03
  = 2.22 m/s (case a).

6.7 Four-bar or quadric cycle chain

As explained in Section 6.3, this is the simplest of the mechanisms and by fixing each link in turn, inversions of the mechanism are obtained.

Fig. 6.9 (a) shows a four-bar chain in which the link *AD* is fixed, i.e. points *A* and *D* become fixed points on the frame of the machine about which links *AB* and *CD* can rotate. Since *A* and *D* are fixed they have no relative velocity. They will thus be represented on the velocity diagram (Fig. 6.9 (b)) by the same point, namely point *a, d*.

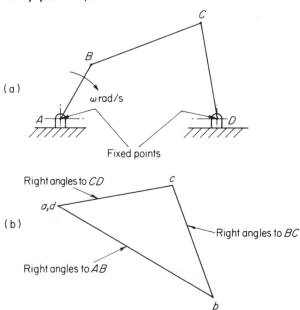

Fig. 6.9 (a) Space diagram
(b) Velocity diagram

It is always desirable to identify all the fixed points on the mechanism at the outset and to represent them by a single point on the velocity diagram. Note that points on the velocity diagram are identified with the lower-case letters corresponding to the capital letters on the mechanism or space diagram. Link *AB* is shown as the driving member in this case, and has a clockwise velocity of ω rad/s.

$$\therefore \text{ Linear velocity of } B \text{ relative to } A = \omega AB.$$

Vector **ab** is drawn to scale to represent this velocity which is at right angles to *AB*.

The velocity of *C* relative to *B* is at right angles to *BC*, so that through point *b* on the velocity diagram a line in this direction is drawn.

The velocity of *C* relative to *D* is at right angles to *CD*, so that through point *d* on the velocity diagram a line in this direction is drawn to intersect the line from *b* at point *c*.

bc defines the linear velocity of *C* relative to *B*
dc defines the linear velocity of *C* relative to *D*.

The simple procedure for constructing the velocity diagram is thus:
  (i) draw the mechanism to scale in the instantaneous position requiring investigation;
  (ii) calculate, if necessary, the input linear velocity of the end of one link;
  (iii) locate all fixed points as one point on the velocity diagram;
  (iv) for each moving link there will then be a vector on the velocity diagram at right angles to the line joining two identified points on a link.

Example 6.3
A mechanism consists of 4 links connected by pin joints. Their lengths are: $AB = 0.2$ m; $BC = 1.2$ m; $CD = 0.5$ m; $AD = 1.4$ m. Link *AB* is driven at 60 rev/min clockwise and link *AD* is fixed. At the instant when link *AB* is vertical with *B* above *A* determine:
  (a) the angular velocity of the link *CD*;
  (b) the linear velocity, relative to *A*, of the mid-point *E* of link *BC*.

Fig. 6.10(a) shows the mechanism drawn to scale in the position described.
Fig. 6.10(b) shows the velocity diagram for this position. The first vector, **ab**, is drawn at right angles to *AB* and is equal in length to 1.26 m/s.
The linear velocity of *B* relative *A* being $= \omega AB$. $\omega = 60$ rev/min and $AB = 0.2$ m.

$$\therefore V_{ba} = 60 \times \frac{2\pi}{60} \times 0.2 = 1.26 \text{ m/s}.$$

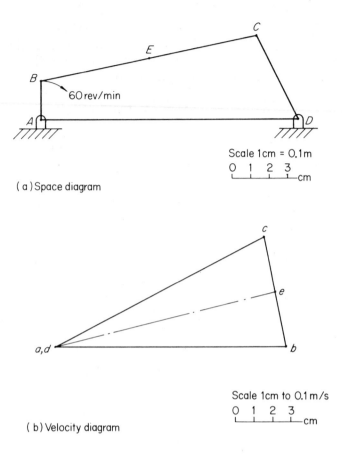

Fig. 6.10 (a) Space diagram
(b) Velocity diagram

The point e on vector **bc** divides the vector in the same proportion as E divides the link BC, i.e. in this case in half.

(i) **cd** is measured, and scales 12.7 cm.

∴ velocity of C relative to D = 12.7 × 0.1 = 1.27 m/s.

Angular velocity of link CD $= \dfrac{cd}{CD} = \dfrac{1.27}{0.5} = 2.5$ rad/s or 23.9 rev/min.

(ii) The velocity of E, the mid-point of BC relative to A, is obtained by joining the point e on the vector diagram to point a and scaling.
ae scales 12.3 cm

∴ velocity of E relative to A = 12.3 × 0.1 = 1.23 m/s.

Example 6.4
A four-bar chain, ABCD, has AD 0.5 m long and fixed. The driving link CD, which is 0.12 m long, rotates at 150 rev/min clockwise. AB is 0.15 m and BC is 0.5 m long. E is a point on a solid extension of BC and is 0.2 m beyond B. For the position when CD is 30° clockwise from the vertical and below D, determine:

Mechanisms and Velocity Diagrams   133

(a) the angular velocities of links *AB* and *BC*;
(b) the linear velocity, relative to *A* of the point *E*;
(c) the linear velocity of *E* relative to *C*;
(d) the rubbing velocity on the pin at *B* if it is 12 mm diameter.

Linear velocity of *C* relative to *D*:

$$\omega CD = \left(150 \times \frac{2\pi}{60}\right) \times 0.12 = 1.9 \text{ m/s}.$$

∴ **cd** is drawn at right angles to *CD* and to scale to represent 1.9 m/s. **cb** is at right angles to *CB* and **ab** is at right angles to *AB*.

Since $\dfrac{BE}{BC} = \dfrac{0.2}{0.5} = 0.4$ then $\dfrac{be}{bc} = 0.4$.

Now **bc** scales 6.3 cm, so that **be** = 6.3 × 0.4 = 2.5 cm.

From the vector diagram Fig. 6.11 (b):
(a) **bc** scales 6.3 cm

∴ Linear velocity of *B* relative to *C* = 6.3 × 0.15 = 0.95 m/s.

From $\omega = \dfrac{v}{r}$,

angular velocity of $BC = \dfrac{bc}{BC} = \dfrac{0.95}{0.5} = 1.9$ rad/s

$= 18.1$ rev/min.

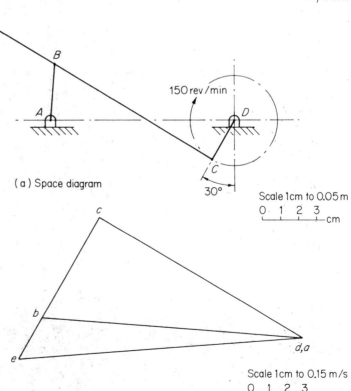

(a) Space diagram

(b) Vector diagram

Fig. 6.11 (a) Space diagram
(b) Velocity diagram

From the scaled length of **ab**, 14 cm, the linear velocity of B relative to $A = 14 \times 0.15$ or 2.1 m/s.

$$\therefore \text{ angular velocity of } AB = \frac{ab}{AB} = \frac{2.1}{0.15} = 14 \text{ rad/s}$$
$$= 134 \text{ rev/min.}$$

(b) **ae** scales 15.3 cm.

$\therefore$ The linear velocity of E relative to $A = 15 \times 0.15 = 2.3$ m/s.

(c) **ce** scales 8.8 cm.

$\therefore$ The linear velocity of E relative to $C = 8.8 \times 0.15 = 1.32$ m/s.

Alternatively: the angular velocity of BE is the same as that of BC; they are the same link.

$\therefore$ Velocity of E relative to $C = 1.9 \times 0.7 = 1.33$ m/s.

(d) Considering the pin at B. Since C is moving anticlockwise relative to B ($b \rightarrow c$ on the velocity diagram), A is moving clockwise relative to B ($b \rightarrow a$ on the velocity diagram), then:

$$\text{rubbing velocity} = (1.9 + 14.0) \times 0.006 = 0.096 \text{ m/s}$$

where the angular velocity of BC is 1.9 rad/s, the angular velocity of AB is 14 rad/s and the pin radius is 0.006 m.

The point E would be used for adding further links to the mechanism.

6.8 Slider-crank chain

This is an inversion of the four-bar chain in which one of the turning pairs is replaced by a sliding pair. Referring to Fig. 6.12:
  (i) AD is the fixed link, represented by two points on the machine frame.
  (ii) AB and BC are rigid links pivoted at A, B and C.
  (iii) C also identifies the slide to which BC is pinned.
  (iv) C moves in a predetermined way relative to D, i.e. along the slide surface.
  (v) The point D is identified by the position of the slide at any instant.

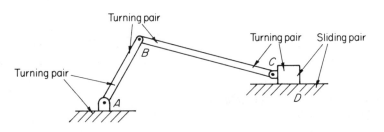

Fig. 6.12 Slider crank chain

It is obvious that the crank and connecting-rod mechanism mentioned in Section 6.1 falls into this category. The slide would be the cylinder, which is stationary, the sliding member would be the piston pinned by the gudgeon pin to the connecting rod at C. The connecting rod is fixed to the crankshaft at B and the crankshaft rotates about A. The throw of the crank being AB.

Once again the velocity diagram is very simple to construct once the instantaneous position of the mechanism is decided. Referring to Fig. 6.13, if the angular velocity of the crank is ω rad/s clockwise then:

(i) linear velocity of $B = \omega AB$;
(ii) points A and D on the mechanism are fixed so that **ad** is a common point on the velocity diagram;
(iii) **ab** is drawn to scale at right angles to AB to represent the velocity of B relative to A, i.e. ωAB;
(iv) C must move at right angles to BC so that through b a line is drawn in this direction;
(v) C must move parallel to D (it slides along the surface) so through d a line is drawn parallel to the slide to intersect the line from b at c;
(vi) **bc** then represents the velocity of C relative to B;
(vii) **dc** then represents the velocity of C relative to D, the sliding velocity.

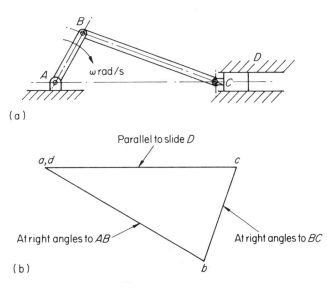

Fig. 6.13 (a) Space diagram
(b) Velocity diagram

Example 6.5
In a crank and connecting-rod mechanism, the crank AB is 10 cm long and the connecting rod BC is 50 cm long. The piston is moving at 4 m/s towards A when the crank is 30° clockwise from the inner dead-centre position.

Determine:
(a) the angular velocity of the crank;
(b) the rubbing velocity at the crank pin if its diameter is 5 cm.

Fig. 6.14 shows the definition of 'inner' and 'outer' dead-centre positions for the crank and connecting rod. The mechanism is now drawn to scale Fig. 6.15 (a).

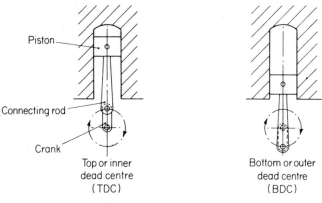

Fig. 6.14 Inner and outer dead-centre positions for a crank and connecting rod

Top or inner dead centre (TDC)

Bottom or outer dead centre (BDC)

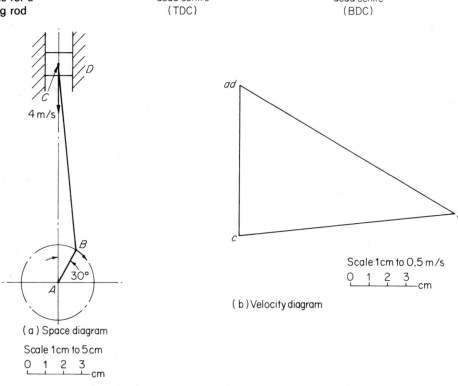

(a) Space diagram
Scale 1cm to 5cm

(b) Velocity diagram
Scale 1cm to 0.5 m/s

Fig. 6.15 (a) Space diagram (b) Velocity diagram

For the velocity diagram, **dc** is drawn first to represent the piston velocity which is given as 4 m/s. Since the piston moves vertically in the cylinder **dc** is vertical with *c* below *d* as the piston is moving downwards having passed through inner dead centre.
cb is at right angles to *CB*, and **ab** is at right angles to *AB*.

(a) **ab** scales 13.3 cm.

∴ Linear velocity of B relative to $A = 13.3 \times 0.5 = 6.65$ m/s.

Angular velocity of crank $AB = \dfrac{ab}{AB} = \dfrac{6.65}{0.1} = 66.5$ rad/s

or 635 rev/min.

(b) **bc** scales 11.7 cm.

∴ Linear velocity of B relative to $C = 11.7 \times 0.5 = 5.85$ m/s.

Angular velocity of $BC = \dfrac{bc}{BC} = \dfrac{5.85}{0.5} = 11.7$ rad/s.

Since BC rotates anticlockwise and AB rotates clockwise about B then the rubbing velocity at the crank pin $= (11.5 + 66.5)\,0.025 = 1.96$ m/s.

Note. In this example the piston was shown moving vertically, which is very often the case. However whichever way the mechanism had been drawn, the relative velocities would have been unaffected.

6.9 Inversions of the slider-crank chain

A very common inversion of the basic mechanism is to allow the slide to move and to fix one of the links. This is the basis of the slow traverse and quick-return mechanism used in a number of machines.

Referring to Fig. 6.16, the crank CD rotates about the fixed point D and the slotted link AE is pivoted at the fixed point A. The trunnion block which slides in the slotted link forms a sliding pair CB. The block is also pinned to end C of the crank to form a turning pair. The slotted link and trunnion block are illustrated by the enlarged scrap view in Fig. 6.17. This is only one of several designs.

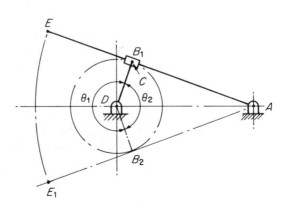

Fig. 6.16 Quick return mechanism

Fig. 6.17 Slotted link and trunnion block

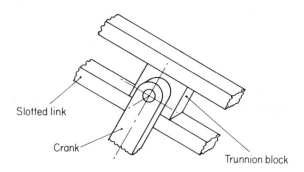

As the crank rotates once, from the position shown, the end E of the slotted link moves from E to E' and back again. The crank is shown rotating anticlockwise so the movement from E to E' takes longer than the movement from E' to E since angle θ_1 is greater than angle θ_2. The point E would be attached by a further link to a machine slide which would thus have different times for inward and outward strokes. A complete mechanism of this kind is illustrated by Example 6.7. First, however, let us look at a numerical example of the basic mechanism.

Example 6.6
In a mechanism of the type illustrated in Fig. 6.16 $AD = 0.6$ m; $DC = 0.2$ m; $AE = 0.8$ m and the crank DC rotates anticlockwise at 143 rev/min. For the instant when the crank is 30° above the inner horizontal position, determine:
(a) the sliding velocity of the trunnion block within the link;
(b) the linear velocity of the point E on the slotted link.

The mechanism is drawn to scale in the position given Fig. 6.18(a).
The linear velocity of C relative to D is calculated.

$$V_{cd} = \omega CD = \left(143 \times \frac{2\pi}{60}\right) \times 0.2 = 3 \text{ m/s}$$

In the velocity diagram, Fig. 6.18(b), the vector **cd** is drawn at right angles to the crank DC and to scale to represent 3 m/s. The vector **cb** is drawn parallel to the slotted link and represents the sliding velocity of C on the trunnion block, relative to B on the slotted link. The vector **ab**, at right angles to the slotted link ABE, represents the linear velocity of B relative to A.
(a) **cb** scales 6.9 cm.

∴ sliding velocity of the trunnion block = $6.9 \times 0.3 = 2.07$ m/s.

(b) To find the linear velocity of E we must first scale the distance AB and the vector **ab**. Then:

$$\frac{AB}{AE} = \frac{ab}{ae}$$

Mechanisms and Velocity Diagrams

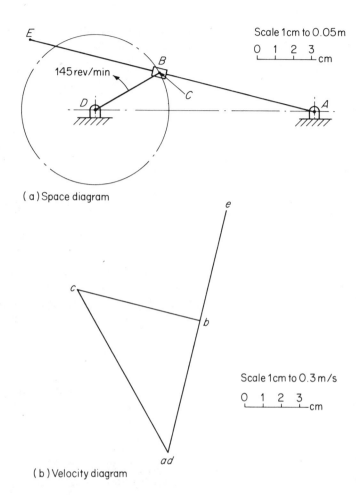

Fig. 6.18 (a) Space diagram
(b) Velocity diagram

ab = 7.3 cm; AE = 16 cm and AB = 8.8 cm.

$$\therefore \mathbf{ae} = \mathbf{ab} \times \frac{AE}{AB} = 7.3 \times \frac{16}{8.8} = 13.3 \text{ cm.}$$

Thus the velocity of $E = 13.3 \times 0.3 = 3.99$ m/s.

It would also be informative to determine the time for the two strokes from E to E' and from E' to E. From Fig. 6.19, which shows

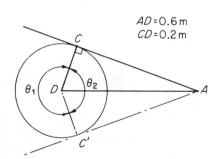

Fig. 6.19 Part of the quick return mechanism

part of the mechanism and AE in one extreme position at right angles to the crank,

$$\cos \angle CDA = \frac{0.2}{0.6} \quad \therefore \angle CDA = 70.5° = \tfrac{1}{2}\theta_2$$

$$\therefore \theta_2 = 141°$$
$$\theta_1 = 360 - 141 = 219°$$
$$\text{Time for 1 rev} = \frac{2\pi}{\omega} = \frac{2\pi}{15} = 0.42 \text{ s}$$

$$\therefore E \text{ to } E' \text{ takes } 0.42 \times \frac{219}{360} \text{ s} = 0.25 \text{ s}$$

$$E' \text{ to } E \text{ takes } 0.42 \times \frac{141}{360} \text{ s} = 0.17 \text{ s}$$

Having now analysed the basic mechanism we can now add a link and a further slide to demonstrate the complete set up in a machine. It should be noted that by doing so we have now moved away from the basic quadric chain.

Example 6.7
In the quick return mechanism shown in Fig. 6.20 (a) the dimensions are: $AD = 0.6$ m; $CD = 0.2$; $AE = 1.2$ m; $x = 0.6$ m and $EF = 0.2$ m. For the crank in the position shown, 45° from the outer horizontal position, determine the velocity with which the machine table is moving. The crank rotates at 76 rev/min anticlockwise.

In the velocity diagram, Fig. 6.20 (b) note that points *a*, *d* and *g* are common as points *A*, *D* and *G* on the mechanism are fixed.

Linear velocity of $C = \left(76 \times \frac{2\pi}{60}\right) \times 0.2 = 1.6$ m/s which is vector **dc** to scale. Length $AB = 7.6$ cm, equivalent to 0.76 m. Vector **ab** scales 6.5 cm;

$$\therefore \text{vector } \mathbf{ae} = \mathbf{ab} \times \frac{AE}{AB} = 6.5 \times \frac{1.2}{0.76} = 10.3 \text{ cm}.$$

Vector **ef** is at right angles to *EF* and vector **fg** is parallel to the slide and passes through *g*. **fg** to scale is the table velocity. **fg** scales 10.3 cm:

$$\therefore \text{table velocity} = 10.3 \times 0.2 = 2.1 \text{ m/s}.$$

6.10 Summary

(i) *Link* – one element in a machine.
(ii) *Pair* – two elements joined.
(iii) *Chain* – a number of pairs joined.

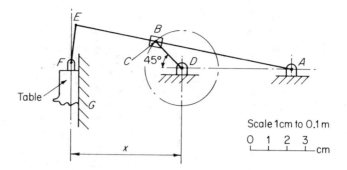

(a) Space diagram

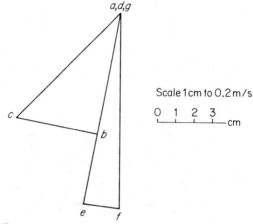

(b) Velocity diagram

Fig. 6.20 (a) Space diagram (b) Velocity diagram

   (iv) *Mechanism* – a number of pairs joined with one link fixed inorder to produce a predictable motion from the remainder.
   (v) *Quadric chain* – the simplest mechanism, consisting of 4 bars or links.
   (vi) *Slider crank chain* – an inversion of the quadric chain with one sliding pair.
   (vii) *Velocity diagrams* represent the instantaneous linear relative velocity of different points in the mechanism.

Exercises 6

Questions 1 to 5 refer to the theory and should be attempted initially without further reference to the text.
  1 In connection with a mechanism indicate which of the following statements are correct.
    (a) A link is one element in a mechanism.

continued

(b) A pair is defined as two links joined together by a pin.
(c) A mechanism is a chain with one link fixed.
(d) The simplest mechanism has four pairs.
(e) The simplest mechanism can have at least three inversions.

2. With reference to the velocities of the various parts of a mechanism indicate which of the following statements are correct.
   (a) All velocities are relative.
   (b) Two links moving in the same direction in a mechanism have a relative velocity equal to the difference between their velocities.
   (c) Each point on a rotating solid link has a different velocity relative to a fixed point.
   (d) Each point on a rotating solid link has the same velocity relative to all fixed points in the mechanism.
   (e) There is no relative velocity between two points on the same solid link.

3. The following statements refer to the rubbing velocity on the pins joining links in a mechanism. Indicate which are correct.
   (a) The rubbing velocity depends upon the angular velocity of the links which the pins join.
   (b) The rubbing velocity depends upon the pin diameter.
   (c) For links moving in the same direction the rubbing velocity at the pin is the sum of their angular velocities.
   (d) For links moving in opposite directions the rubbing velocity is based on the difference between their angular velocities.
   (e) The rubbing velocity will affect the pin wear.

4. In connection with the four-bar chain indicate which of the following statements are correct.
   (a) The quadric-cycle chain is a four-bar chain.
   (b) Two points in the mechanism will have no velocity relative to each other.
   (c) The velocity diagram will always be a triangle.
   (d) Each line will be at right angles to a link of the chain.
   (e) The velocity diagram represents angular velocities.

5. For the single slider-crank chain which of the following statements are correct?
   (a) It is an inversion of the four-bar chain.
   (b) The quick-return mechanism is an inversion of the single slider-crank chain.
   (c) The point where the connecting rod joins the slide has two instantaneous motions.

(d) A trunnion block is a slotted link comprising a sliding pair and a turning pair.
(e) An internal combustion engine usually contains this mechanism.

6 Links AB and BC in a mechanism are joined by a 25 mm diameter pin. Link AB is 0.2 m long and BC is 0.15 m long. For AB the velocity of A relative to B is 2.5 m/s and for BC the velocity of C relative to B is 4.2 m/s. If the two links have an instantaneous rotation about B determine:
  (i) their angular velocities;
  (ii) the rubbing velocity on the pin at B if the links are rotating in (a) in the same direction; (b) in opposite directions.

7 In an engine the crank pin is 105 mm diameter and the gudgeon-pin is 20 mm diameter. The instantaneous velocity of the crank is 150 rad/s clockwise about the crank pin. The connecting rod is 0.3 m long measured between pin centres. The relative linear velocity between the pin centres on the connecting rod is 6 m/s and the rod is moving anticlockwise about the gudgeon pin and clockwise about the crank pin. Determine:
  (i) the rubbing velocity at the crank pin;
  (ii) rubbing velocity at the gudgeon pin.

8 In a four bar chain $ABCD$, link AD is horizontal and fixed and is 0.5 m long. $AB = 0.2$ m; $BC = 0.4$ m; $CD = 0.3$ m. AB rotates clockwise about A at 544 rev/min. For the instant when the crank AB is 45° above the horizontal and closest to D determine:
  (i) the angular velocity of each link;
  (ii) the linear velocity of the mid-point of BC relative to A.

9 A mechanism consists of four links AB, BC, CD and DA connected by pin joints. Link AB is 0.25 m long and fixed in the vertical position with B above A. $BC = 0.15$ m; $CD = 0.4$ m; and $AD = 0.25$ m. The mechanism is driven by BC which rotates at a constant velocity of 1 200 rev/min clockwise about B. For the instant when BC is vertical with C above B determine:
  (i) the linear velocity of D relative to C;
  (ii) the angular velocity of CD and AD;
  (iii) the rubbing velocity on the pin at C if it is 20 mm diameter.

10 A quadric cycle chain $ABCD$ has the link CD fixed horizontally; it is 200 mm long. AD, which is 100 mm long, is driven anticlockwise at 480 rev/min. $AB = 180$ mm;

continued

BC = 150 mm. BE is a solid extension of CB such that BE = 40 mm. For the instant when AD has moved 60° from the vertical position with A below D determine:
  (i) the angular velocity of AB and BC;
  (ii) the linear velocity of E relative to C;
  (iii) the linear velocity of E relative to A.

11 In an engine, the crank is 100 mm long and the connecting rod is 300 mm long. The centre of rotation of the crank lies on the vertical centreline of the piston which moves vertically. The piston is above the crank. At a certain instant the crank is 40° clockwise from the inner dead-centre position. Determine:
  (i) the piston velocity;
  (ii) the angular velocity of the connecting rod if the crank rotates at 600 rev/min clockwise.

12 Referring to Fig. 6.12, in a single slider-crank chain the centre of the crank is 0.1 m below the horizontal centreline of the sliding member. The crank AB is 0.2 m long and the connecting rod BC 0.8 m long. The crank is driven at a constant velocity of 120 rev/min clockwise. For the instant when AB is 30° clockwise from the vertical, B above A, determine:
  (i) the velocity of the slide;
  (ii) the velocity of the mid-point of the connecting rod relative to A;
  (iii) the rubbing velocity on the crank pin if it is 50 mm diameter.

13 In a quick-return mechanism, similar to that shown in Fig. 6.16, the crank DC is 0.2 m long. The distance AD is 1.0 m and the link AE is 1.5 m long. For a position when the crank is vertical, above AD, determine:
  (i) the linear velocity of E relative to A;
  (ii) the linear velocity of the trunnion block along AE; if the crank rotates clockwise at 60 rev/min.
Determine also the maximum angular displacement of AE.

14 In the mechanism shown in Fig. 6.21 the distance AD is

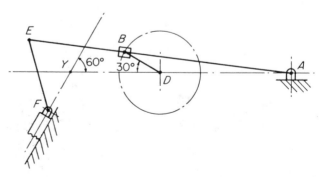

Fig. 6.21 The space diagram of the mechanism in question 14

2.0 m. The crank is 0.5 m long and rotates clockwise at 300 rev/min. The driving link AE is 3.5 m and EF is 1.0 m long. The slide, joined to EF at F, moves at 60° from the horizontal as shown. DY = 1 m.
(a) For the position of the mechanism shown when angle θ = 30° determine:
  (i) the velocity of the slide;
  (ii) the velocity of the trunnion along AE.
(b) Determine the linear distance moved by the slide in one stroke.

15 For a mechanism similar to that shown in Fig. 6.20 if AD = 0.8 m, x = 0.6 m, AE = 1.4 m and EF = 0.3 m determine for a position when the crank is:
  (i) vertical with C above D;
  (ii) vertical with C below D;
the linear velocity of the table. The crank is 0.25 m long and rotates anticlockwise at 240 rev/min.

7 Flywheels and Turning Moment Diagrams

7.1 Introduction

The main function of a flywheel is to act as a reservoir of energy in a machine. It is usually interposed between the driving and driven member and it is required either because the driving member is unable to maintain a constant supply of energy or the driven member offers a varying resistance to motion. Sometimes both the supply and the resistance may vary. Without the energy supplied by the flywheel large and intolerable fluctuations in speed would occur.

The internal combustion engine using the crank and connecting-rod mechanism is an example of the varying supply of energy. Only during the power or expansion stroke is energy supplied; during the exhaust, induction and compression strokes energy is extracted. The crank press is an example of a varying load or resistance, the maximum resistance occurs at the instant of punching or forming, with only frictional resistance during the remainder of the cycle.

To achieve its main function efficiently, therefore, the flywheel must keep the speed variation to a desired minimum and at the same time it must impose a minimum load on the machine. The conditions under which it will operate must be carefully investigated.

7.2 Flywheel energy

This is essentially *kinetic energy* or energy of motion.

$$\text{Kinetic energy or KE} = \frac{1}{2}I\omega^2$$

where I = moment of inertia = mass × (radius of gyration)^2
 = mk^2

ω = angular velocity in rad/s.

For a solid disc flywheel $k^2 = \dfrac{r^2}{2}$ where r is the outside radius

For the rim or annular type $k^2 = \dfrac{r_1^2 + r_2^2}{2}$ where r_1 and r_2 are the inner and outer radii.

Example 7.1
A solid disc flywheel 0.4 m diameter and 10 cm thick is made from cast iron of density 7 Mg/m³. Determine its moment of inertia.

$$I = mk^2$$

$$m = \text{volume} \times \text{density} = \pi r^2 t \rho$$
$$= \pi \times 0.2^2 \times 0.1 \times 7 \times 10^3$$
$$= 88 \text{ kg}$$

$$k^2 = \frac{0.2^2}{2} = 0.02 \text{ m}^2$$

$$\therefore \quad I = 88 \times 0.02 = 1.76 \text{ kg m}^2.$$

Example 7.2
For the flywheel in Example 7.1 calculate the kinetic energy at 1000 rev/min.

$$KE = \frac{1}{2} I \omega^2$$

where $I = 1.76$ kg m²

$$\omega = 1000 \times \frac{\pi}{30} = 33.3\pi \text{ rad/s}$$

$$\therefore \quad KE = \frac{1}{2} \times 1.76 \times (33.3\pi)^2 = 9650 \text{ J}$$

Example 7.3
The speed of the flywheel on a shearing machine is initially 120 rev/min. During the shearing operation, which takes 2 seconds and requires 22 000 J of energy, the speed must not fall below 90 rev/min. The driving motor supplies 3 kW at 105 rev/min. The loss of energy due to friction is 25% of that supplied. Determine:
 (a) the moment of inertia of the flywheel;
 (b) its radius of gyration if its mass is 650 kg.

Energy supplied by the motor in 2 s = 3000 × 2 = 6000 J
Energy to friction in 2 s = 6000 × 0.25 = 1500 J
Energy available for shearing = 6000 − 1500 = 4500 J
Energy required for shearing = 22 000 J
$\therefore$ Energy from the flywheel = 22 000 − 4500 = 17 500 J

Initial energy of flywheel = $\dfrac{1}{2} I \omega_1^2$

Final energy of flywheel $= \frac{1}{2} I \omega_2^2$

where $\omega_1 = 120$ rev/min $= 120 \times \frac{\pi}{30} = 4\pi$ rad/s

$\omega_2 = 90$ rev/min $= 90 \times \frac{\pi}{30} = 3\pi$ rad/s

Loss of energy of flywheel $= \frac{1}{2} I(\omega_1^2 - \omega_2^2) = 17\,500$ J.

$$\therefore I = \frac{2 \times 17\,500}{(4\pi)^2 - (3\pi)^2} = \frac{35\,000}{7\pi^2}\text{ kg m}^2$$

$$= 506.6 \text{ kg m}^2$$

$$k^2 = \frac{I}{m} = \frac{506.6}{650} \text{ m}^2$$

$$k = \sqrt{0.78} = 0.883 \text{ m}$$

7.3 Turning moment diagrams

Designing a flywheel to suit the requirements of a particular machine, requires studying the cyclic variation in
 (i) the torque from the driving member; or
 (ii) the resisting torque offered by the driven member; or
 (iii) both of these.
The torque variation is plotted vertically to a base of revolutions, crank angle or time and is known as the *turning moment diagram*.

Fig. 7.1 shows the turning moment developed by a single cylinder engine for one complete cycle. Only on the power stroke does the engine provide a driving or positive torque. During the exhaust, induction and compression strokes, energy is required to

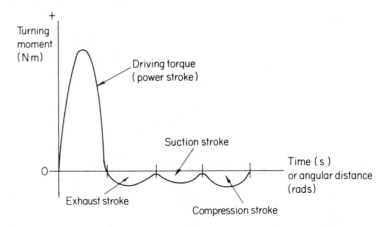

Fig. 7.1 Turning moment diagram for a single cylinder engine

move the piston in the cylinder. It is during this latter period that the engine would slow down considerably as kinetic energy is converted into the work energy to provide these movements. The flywheel must be designed to keep the speed variation to a specified minimum.

Fig. 7.2 shows the resisting torque or turning moment from a crank press or shearing machine during one cycle. The sudden peak represents the shearing operation which lasts for only a fraction of the cycle. It would result in a sudden demand on the driving motor, a dramatic drop in speed and a considerable overload but for the provision of a flywheel which provides the reserve of energy.

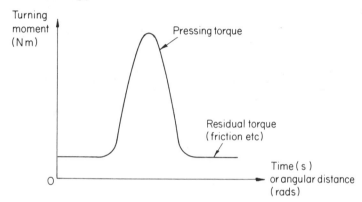

Fig. 7.2 Turning moment diagram for a crank press

Mean torque

If the turning moment diagram is plotted to a base of angle turned in radians then the area of the diagram above and below the base represents work done or work energy. Areas above the base represent positive values and those below represent negative values. Then:

$$\text{Mean torque } (T_m) = \frac{\text{energy supplied or absorbed/cycle}}{\text{angular displacement/cycle}}$$

For the engine, the mean torque is usually assumed to be the resisting torque on the output shaft which for simpler cases is assumed to be constant for a given speed.

For the press or shearing machine the mean torque is assumed to be that provided by the driving motor which runs at approximately constant speed.

In both cases it is the design of the flywheel which limits the variation in the output speed of the engine, or the input speed of the driving motor.

Speed variation

If the mean torque line is inserted on the turning moment diagram for the engine then from Fig. 7.3:

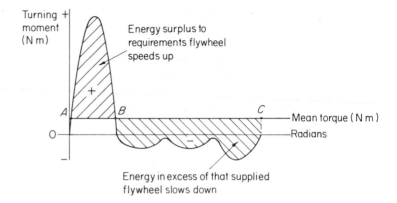

Fig. 7.3 Turning moment and mean torque for an engine

(i) areas above the mean torque line represent energy in excess of requirements at the output shaft;
(ii) areas below the mean torque line represent energy requirements, provided by the flywheel, to supplement those from the engine.

Thus if the flywheel is rotating at ω_1 rad/s at A then at B it will have speeded up due to the surplus energy absorbed and have achieved a maximum velocity ω_2. From B to C energy is extracted and the wheel slows so that at C the speed is ω_1 again and the cycle repeats. The areas above and below the mean torque line for one cycle are obviously equal.

If the mean torque line is inserted on the turning moment diagram for the press then from Fig. 7.4:

(i) areas above the mean torque line represent energy demands above those provided by the driving motor;
(ii) areas below the mean torque line represent energy surplus to requirements.

Thus at A (Fig. 7.4) if the flywheel is rotating at ω_1 rad/s then at B it will have slowed to ω_2 rad/s having supplied energy to the pressing operation. During the remainder of the cycle, from B to C the energy is regained as the flywheel speeds up again.

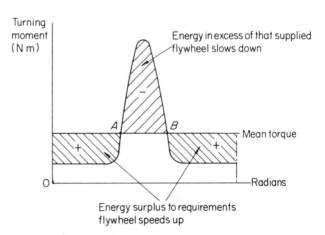

Fig. 7.4 Turning moment and mean torque for a crank press

Example 7.4

The turning moment diagram for an engine was drawn on a base of crank angle. The mean torque line was added and the areas above and below the line were measured. They were in sequence: $+5.5$; -1.5; $+1.7$; and -5.7 cm². The scales used were 1 cm to 1000 N m for the torque axis and 1 cm to 15° for the crank angle. Determine the mass of a flywheel to keep the speed between 295 and 305 rev/min if its radius of gyration is 0.5 m.

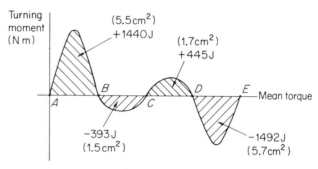

Fig. 7.5 The turning moment diagram

On the turning moment diagram $1 \text{ cm}^2 = 1000 \text{ Nm} \times \dfrac{15° \text{ rad}}{57.3}$

$= 262 \text{ J}$

∴ 5.5 cm² = 1440 J; 1.5 cm² = 393 J; 1.7 cm² = 445 J; 5.7 cm² = 1492 J.

Referring to Fig. 7.5:
  surplus energy at $A = 0$;
  surplus energy at $B = 1440$ J;
  surplus energy at $C = 1440 - 393 = 1047$ J;
  surplus energy at $D = 1047 + 445 = 1492$ J;
  surplus energy at $E = 1492 - 1492 = 0$.

The speed is thus maximum at D and minimum at E and A.
The maximum surplus of energy which is called the maximum *fluctuation of energy* (E_F), is 1492 J.

$$E_F = \frac{1}{2}I(\omega_1^2 - \omega_2^2)$$

where $\omega_1 = 305$ rev/min $= 31.93$ rad/s
  $\omega_2 = 295$ rev/min $= 30.89$ rad/s
  $E_F = 1492$ J

$$\therefore \quad 1492 = \frac{1}{2}I(31.93^2 - 30.89^2)$$

$$I = \frac{2 \times 1492}{66} = 45.2 \text{ kg m}^2$$

$$m = \frac{I}{k^2} = \frac{45.2}{0.5^2} = 180.8 \text{ kg}$$

Example 7.5

A motor driving a punching machine exerts a constant torque of 675 N m on a flywheel which rotates at an average speed of 120 rev/min. The punch operates 60 times per minute and the actual punching operation takes 0.2 seconds. Assuming that the punching operation requires a constant torque and neglecting friction, determine the punching torque. If the flywheel speed must not vary by more than 10 rev/min determine the moment of inertia of the flywheel.

Sketch the turning moment diagram on a base of flywheel angle and indicate where the maximum and minimum speeds will occur.

To determine the flywheel displacement for 1 cycle:
time for one complete cycle = 60/60 = 1 s;

$$\text{flywheel velocity} = 120 \times \frac{\pi}{30} = 4\pi \text{ rad/s};$$

angular displacement for 1 cycle = 4π radians;
time for punching = 1/5 s;

$$\text{angular displacement in } 1/5 \text{ s} = \frac{4\pi}{5} = 0.8\pi \text{ radians};$$

Referring to Fig. 7.6, since friction is neglected:

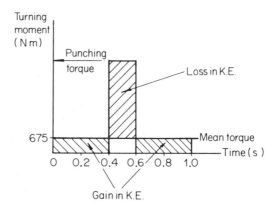

Fig. 7.6 The outline turning moment diagram for the punching machine

energy for punching in 0.2 s (0.8π rads) = energy supplied in 1 s. (4π rads)

$$\therefore T_{max} \times 0.8\pi = T_{mean} \times 4\pi$$

$$\therefore T_{max} = T_{mean} \times \frac{4\pi}{0.8\pi} = 675 \times \frac{4}{0.8} \text{ N m}.$$

$$= 3375 \text{ N m}.$$

Loss in kinetic energy during punching = area above mean torque line (Fig. 7.6)

$$(T_{max} - T_{mean})0.8\pi = (3375 - 675)0.8\pi = 6786 \text{ J}$$

The loss in kinetic energy can also be written in the form:

$$KE = \frac{1}{2}I(\omega_1^2 - \omega_2^2)$$

where I is the required moment of inertia and ω_1 and ω_2 are the maximum and minimum speeds.

The mean speed is 120 rev/min. This must not vary by more than 10 rev/min i.e. ± 5 rev/min.

$$\omega_1 = 125 \text{ rev/min} = 125 \times \frac{\pi}{30} \text{ rad/s} \quad \therefore \quad \omega_1^2 = 171.3$$

$$\omega_2 = 115 \text{ rev/min} = 115 \times \frac{\pi}{30} \text{ rad/s} \quad \therefore \quad \omega_2^2 = 145$$

$$\therefore \quad \text{Loss in KE} = \frac{1}{2}I(171.3 - 145) = 6786 \text{ J}$$

$$\text{and } I = \frac{2 \times 6786}{26.3} = 516.6 \text{ kg m}^2$$

Fig. 7.7 shows the complete turning moment diagram.

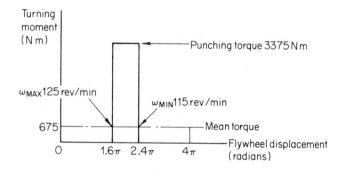

Fig. 7.7 The complete turning moment diagram for the punching machine

7.4 Coefficient of fluctuation of speed (K_s)

The speed variation during a cycle is usually more conveniently expressed as a coefficient which is known as the *coefficient of fluctuation of speed, K_s*.

$$K_s = \frac{\text{maximum speed} - \text{minimum speed}}{\text{mean speed}} = \frac{\omega_1 - \omega_2}{\omega_{\text{mean}}}.$$

Thus in Example 7.5:

$\omega_{\max} = \omega_1 = 125$ rev/min; $\omega_{\min} = \omega_2 = 115$ rev/min; $\omega_{\text{mean}} = 120$ rev/min.

$$\therefore \quad K_s = \frac{125 - 115}{120} = 0.0833.$$

7.5 Coefficient of fluctuation of energy (K_F)

$$K_f = \frac{\text{greatest fluctuation of energy/cycle}}{\text{work done/cycle}} = \frac{E_f}{E}$$

The fluctuation of energy $= \frac{1}{2}I(\omega_1^2 - \omega_2^2)$ and this can be factorised thus: $\frac{1}{2}I(\omega_1 + \omega_2)(\omega_1 - \omega_2)$.

$$\frac{\omega_1 + \omega_2}{2} = \omega_{\text{mean}} \text{ for small speed variations}$$

$$\therefore \quad \omega_1 + \omega_2 = 2\omega_{\text{mean}}$$

From Section 7.4:

$$\omega_1 - \omega_2 = K_s \omega_{\text{mean}}.$$

$$\therefore \quad K_E = \frac{\frac{1}{2}I(2\omega \times K_s\omega)}{E}$$

$\omega = \omega_{\text{mean}}$

$$K_E = \frac{K_s I \omega^2}{E} \quad \text{or} \quad K_s = \frac{K_E E}{I \omega^2}$$

Once again, from Example 7.5:

$K_s = 0.0833$; $I = 516.6$ kg m²; $\omega = 4\pi$ rad/s

$E =$ Mean torque × angular distance
$= 675 \times 4\pi$
$= 8482$ J.

$$\therefore \quad K_E = \frac{0.0833 \times 516.6 \times (4\pi)^2}{8482}$$

$= 0.8$.

The remaining worked examples in this chapter are intended to illustrate the application of the principles previously explained with somewhat more involved problems.

Example 7.6
A multicylinder engine is delivering 300 kW at an average speed of 250 rev/min. The turning moment for one revolution is plotted and the mean torque is determined. The areas above and below the mean torque line are then evaluated as follows:
-0.32; $+3.8$; 0; -2.58; $+3.2$; -2.9; $+2.25$; -3.75; $+0.3$, (all in cm²). The scale used is 1 cm to 5500 N m vertically, and 1 cm to 25° horizontally. The engine flywheel has a mass of 900 kg and a radius of gyration of 0.75 m. Determine the value of:
  (a) the coefficient of fluctuation of speed;
  (b) the coefficient of fluctuation of energy.

Indicate on a sketch of the turning moment diagram the positions of maximum and minimum speed.

From Fig. 7.8 and the information given:

$$1 \text{ cm}^2 = 5500 \times \frac{25\pi}{180} = 2400 \text{ J/cm}^2$$

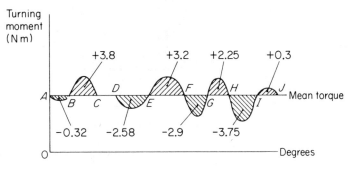

Fig. 7.8 (All areas are in cm²)

Then using areas to represent energy:
  energy at $B = -0.32$;
  energy at $C = -0.32 + 3.8 = +3.48$;
  energy at $D = +3.48 - 0 = +3.48$;
  energy at $E = +3.48 - 2.58 = +0.9$;
  energy at $F = +0.9 + 3.2 = +4.1$;
  energy at $G = +4.1 - 2.9 = +1.2$;
  energy at $H = +1.2 + 2.25 = +3.45$;
  energy at $I = +3.45 - 3.75 = -0.3$;
  energy at $J = -0.3 + 0.3 = 0$.

The minimum energy level is -0.32 at B
The maximum energy level is $+4.1$ at F

These will be the instants of minimum and maximum speed respectively. Greatest energy fluctuation = 4.42 × 2400 J/cm² = 10 608 J or 10.608 kJ.
Work done in one revolution = 2π × mean torque.

$$\text{Mean torque} = \frac{\text{power}}{\text{angular velocity}} = \frac{300 \times 1000}{250 \times \frac{\pi}{30}} = 11\,460 \text{ N m}.$$

∴ Work done in one revolution = $2\pi \times 11\,460 = 72\,000$ J

$$\therefore K_E = \frac{10\,608}{72\,000} = 0.147$$

$$K_s = \frac{K_E E}{I\omega^2} = \frac{0.147 \times 72\,000}{(900 \times 0.75^2) \times \left(250\frac{\pi}{30}\right)^2}$$

$$= \frac{10\,584}{506.25 \times 685.4} = 0.031$$

Example 7.7
The turning moment diagram for a single cylinder double-acting engine may be assumed to consist of two isosceles triangles, each having a base of π radians crank angle. The maximum torque values are 6800 N m and 5400 N m respectively. The flywheel has a radius of gyration of 1 m and the crankshaft rotates at 110 rev/min. Determine:
  (a) the coefficient of fluctuation of energy;
  (b) the flywheel mass necessary to limit the speed fluctuation to $\mp 2\frac{1}{2}$ % of the mean speed.

Referring to Fig. 7.9:

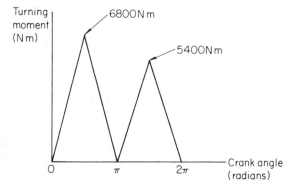

Fig. 7.9 The turning moment diagram for the single cylinder double-acting engine

energy supplied in one revolution (2π radians)
  = area of triangles
$$= \frac{6800\pi}{2} + \frac{5400\pi}{2} = 19\,163 \text{ J}$$

$$\text{Mean torque} = \frac{19\,163}{2\pi} = 3050 \text{ N m}$$

The diagram is now redrawn showing the mean torque line and the peak values above and below that line.
Then referring to Fig. 7.10; by similar triangles:

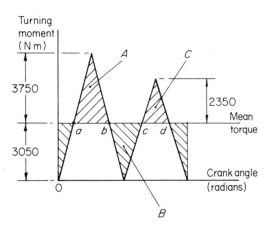

Fig. 7.10 The turning moment diagram showing the mean torque line and peak values above and below that line

$$\frac{ab}{\pi} = \frac{3750}{6800} \text{ thus } ab = 1.73 \text{ rad.}$$

$$\frac{cd}{\pi} = \frac{2350}{5400} \text{ thus } cd = 1.37 \text{ rad.}$$

Then $bc = \pi - \frac{1}{2}(ab + cd) = \pi - \frac{3.1}{2} = 1.59$ rad.

Area of $A = \frac{1.73}{2} \times 3750 = 3244$ J.

Area of $B = \frac{1.59}{2} \times 3050 = 2425$ J (negative).

Area of $C = \frac{1.37}{2} \times 2350 = 1610$ J.

Then taking point *a* as datum:

> energy at $a = 0$
> energy at $b = +3244$ J
> energy at $c = +3244 - 2425 = +719$ J
> energy at $d = +719 + 1610 = +2329$ J

∴ Maximum fluctuation = 0 to 3244 J

$$K_E = \frac{3244}{19\,163} = 0.169$$

Now $K_s = \pm 2\frac{1}{2}\% = 0.05$

Then, from $K_E = \frac{K_s I \omega^2}{E}$:

$$I = \frac{K_E E}{K_s \omega^2}$$

$\omega = 110 \times \frac{\pi}{30} = 11.52$ rad/s

$\omega^2 = 132.7$.

$$\therefore I = \frac{0.169 \times 19\,163}{0.05 \times 132.7}$$

$$= 488 \text{ kg m}^2$$

Since $I = mk^2$ and $k = 1$ m:

$$m = 488 \text{ kg.}$$

Example 7.8

The diagram in Fig. 7.11 shows the variation with time of the torque required on the driving shaft of a punching machine during one cycle of operations. The shaft is directly coupled to a motor

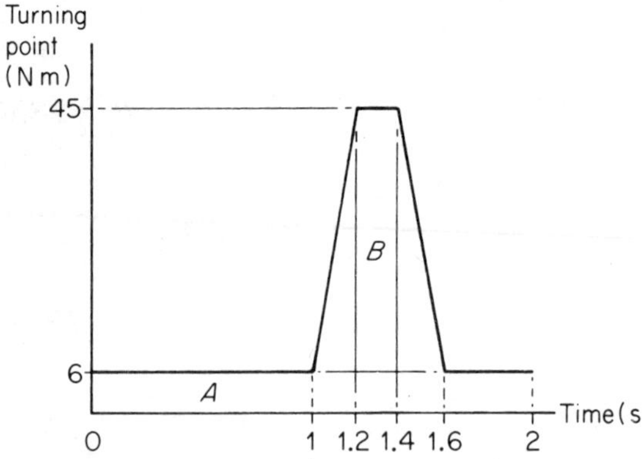

Fig. 7.11 A diagram showing the variation of the torque on the driving shaft of the punching machine during one cycle

which exerts a constant torque and runs at a mean speed of 1200 rev/min. The flywheel attached to the same shaft has a moment of inertia of 1.75 kg m². Determine:
(a) the motor power;
(b) the coefficient of speed fluctuation.

(a) Motor power = mean torque × velocity (rad/s).

$$\text{mean torque} = \frac{\text{area of diagram}}{\text{length of base}} \quad \text{(Fig. 7.11)}.$$

Area $A = 6 \times 2 = 12$; Area $B = \dfrac{(0.6 + 0.2)}{2} \times 39 = 15.6$;

$$\therefore \quad \text{mean torque } (T) = \frac{12 + 15.6}{2} = 13.8 \text{ N m}.$$

$$\text{Power} = T\omega = 13.8 \times 1200 \times \frac{\pi}{30} = 1734 \text{ W or } 1.734 \text{ kW}$$

(b) The diagram is now redrawn to show the areas above and below the mean torque line. The base is converted to a scale in radians since the areas need to represent energy (Fig. 7.12).

$$\text{Velocity in rad/s} = 1200 \times \frac{\pi}{30} = 125.67 \text{ rad/s}$$

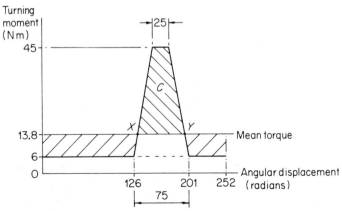

Fig. 7.12 The diagram with the mean torque line inserted

∴ distance in 1 second = 125.67 rad., in 0.2 s = 25 rad. and 1.6 s = 201 rad.

Area C represents the energy extracted from the flywheel during the cycle. By similar figures:

$$\frac{75}{39} = \frac{xy}{45 - 13.8}$$

$$xy = \frac{75 \times 31.2}{39} = 60 \text{ rad.}$$

$$\text{Area of } C = \frac{25 + 60}{2} \times 31.2 = 1326 \text{ J.}$$

∴ Energy extracted = $1326 = \frac{1}{2}I(\omega_1^2 - \omega_2^2)$

$(\omega_1^2 - \omega_2^2) = (\omega_1 - \omega_2)(\omega_1 + \omega_2)$ but $\frac{\omega_1 + \omega_2}{2} = \omega_{mean}$

= 125.67 rad/s

∴ $(\omega_1 + \omega_2) = 2 \times 125.67 = 251.3$ rad/s

Since $I = 1.75$ kg m²,

$$\frac{1}{2}I(\omega_1^2 - \omega_2^2) = \frac{1}{2} \times 1.75 \times 251.3(\omega_1 - \omega_2)$$

$$= 220(\omega_1 - \omega_2)$$

∴ $1326 = 220(\omega_1 - \omega_2)$

$(\omega_1 - \omega_2) = 6.027$

$$K_s = \frac{\omega_1 - \omega_2}{\omega_{mean}} = \frac{6.027}{125.67} = 0.048$$

The design of a flywheel has, therefore, to take into account the fluctuation in both energy and speed per cycle. However, the dimensions and mass of the flywheel may also be constrained by the design of the machine to which it is to be fitted, and to the space available. Since the moment of inertia $I = mk^2$ the radius of gyration k, used intelligently, can economise in the overall mass required. Thin, large diameter, solid discs and rims connected to a hub by spokes are examples of such designs.

7.6 Summary

Flywheels have the main function of providing energy for an engine or machine to keep speed fluctuation to a minimum. For engines they may also assist in the starting process.

$KE = \frac{1}{2}I\omega^2$.

(i) *Energy fluctuation* = $\frac{1}{2}I(\omega_1^2 - \omega_2^2)$.

(ii) *Coefficient of speed fluctuation* $= K_s = \dfrac{K_E E}{I\omega^2}$ or $\dfrac{\omega_1 - \omega_2}{\omega_{mean}}$.

(iii) *Coefficient of energy fluctuation* $= K_E = \dfrac{K_E I \omega^2}{E}$ or $\dfrac{\text{energy fluctuation}}{\text{energy supplied}} \Big/ \text{cycle}$.

(iv) *Turning moment diagram* – shows the cyclic torque variation.

(v) *Mean torque* – for engines the resisting torque on the output shaft;
– for presses etc., the torque applied at the input shaft.

Exercises 7

Questions 1 to 3 concern the theory and should be attempted initially without further reference to the text.

1 The function of a flywheel is:
(a) to reduce the load on shearing or pressing machine motors.
(b) to reduce the speed variation in such motors.
(c) to drive an internal combustion engine at certain periods in the cycle.
(d) to decrease the starting torque for an engine
(e) to act as an energy store.
Indicate the correct statements.

2 The turning moment diagram:
(a) is a graph of torque variation against time.
(b) is a graph of torque variation against crank angle.
(c) shows the resisting torque in a machine press.
(d) shows the driving torque for an engine.
(e) has an area representing the work done during a cycle.
Indicate the correct statements.

3 The coefficients of energy and speed fluctuation are
(a) dependent upon the mean torque.
(b) dependent upon the area of the turning moment diagram.
(c) relate to the flywheel and not the engine or motor.
(d) $\dfrac{\text{energy fluctuation/cycle}}{\text{energy supplied/cycle}}$.
(e) $\dfrac{\text{maximum speed} - \text{minimum speed}}{\text{maximum speed}}$.

Indicate the correct statements.

4 A solid disc flywheel has a moment of inertia of 0.4 kg m². It is 0.3 m diameter and 0.15 m thick. Determine the density of the material from which it is made and its kinetic energy at 1500 rev/min.

5 The flywheel in a machine press has a mass of 700 kg and a radius of gyration of 0.75 m. The punching operation absorbs 30 kJ of energy and friction accounts for 15% of the energy supplied. The driving motor develops 12 kW at 240 rev/min. If the pressing operation takes 1.5 s, determine the speed fluctuation of the flywheel.

6 An engine drives a machine at a mean speed of 210 rev/min. The resisting torque of the machine varies during each revolution of the flywheel as shown in Fig. 7.13. By using a flywheel with a radius of gyration of 0.6 m the coefficient of fluctuation of speed is limited to 0.03. Determine:
  (i) the mean engine power;
  (ii) the mass of the flywheel.

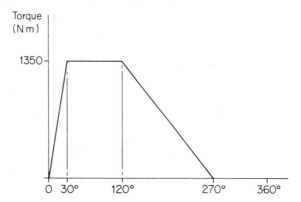

Fig. 7.13 Diagram showing the resisting torque during each revolution

7 The turning moment diagram for an engine is drawn for 1 revolution and the mean resisting torque determined. Areas above and below the mean torque line are measured and the energy variations relative to the line are determined as follows: +2556; −1491; +1491; −1775; +923; −213; +1420; −2911 all in joules. If the engine speed is 2400 rev/min and the fluctuation of speed is not to exceed ±1.2% of the mean speed determine the moment of inertia of the flywheel.

8 Fig. 7.14 shows the turning moment diagram for one complete cycle of a single cylinder four-stroke engine. The engine has an output of 5.5 kW at 750 rev/min. If the maximum torque is 7 times the minimum torque and the resisting torque is constant determine:

continued

(i) the moment of inertia of the flywheel to restrict the speed variation to ±2% of the mean speed;
(ii) the coefficient of energy fluctuation.

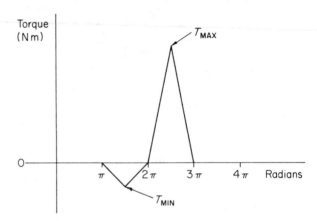

Fig. 7.14 The turning moment diagram for one complete cycle

9 A punching machine which operates 30 times per minute is driven by a motor which supplies a constant torque of 800 Nm at 150 rev/min. The punching torque rises from zero to a maximum in 0.05 s, remains constant for 0.2 s and finally drops to zero in 0.05 s. Neglecting friction determine:
   (i) the punching torque;
   (ii) the moment of inertia of the flywheel if its speed is not to vary by more than ±5%.

10 If the machine in question 9 had a constant frictional torque of 50 N m determine:
   (i) the punching torque;
   (ii) the mass of a wheel of radius of gyration 0.75 m which will keep the speed variation to ±5%;
   (iii) the coefficient of fluctuation of energy.

8 Vibrations

8.1 Introduction

Machines with moving parts must inevitably generate forces which will cause other parts within the machine to vibrate. This in itself, if unchecked, may cause damage or malfunctioning but the vibrations may also be transmitted to foundations and floors causing more general damage. Although it is not possible to eliminate all vibrations, by understanding their nature the worst effects, particularly in cases of resonance, can be avoided; the lesser effects can be minimised.

Vibration may be linear or angular and it can occur in any one of three mutually perpendicular planes. In more complicated cases the vibration may be a combination of instantaneous motions in more than one of these planes. Since for each plane there are two possible types of motion this leads to the concept that a body has six possible degrees of freedom or modes of vibration. In practice however, the actual number of degrees of freedom may be deliberately restricted so that only one is possible.

Fig. 8.1 shows a piston in a cylinder. The piston has only one degree of freedom as it can slide and not rotate within the cylinder. Fig. 8.2 shows a gearwheel mounted on a shaft. The wheel can rotate about one axis only but if the elasticity of the shaft is taken into account it can move linearly in all three planes.

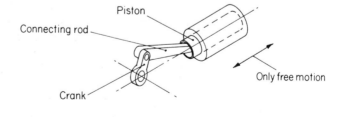

Fig. 8.1 Piston and connecting rod

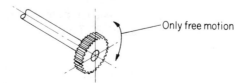

Fig. 8.2 Gearwheel

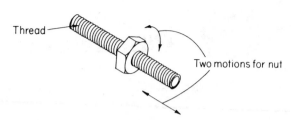

Fig. 8.3 Nut on thread

In both these cases there is only one desired degree of freedom, the others however are unavoidable and may result in vibration. Other components are designed to have more than one degree of freedom. Fig. 8.3 shows a nut on a thread which is obviously designed to have rotation in one plane and linear motion in another.

In the study of vibrations therefore it is first necessary to identify all the possible degrees of freedom of the vibrating body.

8.2 Damping

In practice it will be appreciated that damping in some form will always be present under normal circumstances. This may be due to air resistance, friction, fluid viscosity etc. However in order to develop the theory of *simple vibration*, damping will be ignored, so that, once started, a vibration will continue undiminished.

8.3 Simple vibration

The theory is exactly the same as that for *simple harmonic motion* (SHM) which requires that during the motion of the vibrating body:
(i) its acceleration at any instant is proportional to its displacement from a fixed point in its path;
(ii) the acceleration is directed towards that fixed point.
The motion may be either linear or angular to satisfy these requirements. To establish relationships between displacement, velocity and acceleration, as well as to determine frequency and periodic time, the SHM circle is used.

In Fig. 8.4 the point P rotates with an angular velocity of ω_n rad/s about the point O at a radius of r. Q is the projection of P upon diameter AB and as P rotates completely around the circle Q oscillates (vibrates) between A and B. It is the motion of Q which is *simple vibration* as will be shown.

x is the displacement of Q from centre O t seconds after P has passed B. Thus:

$$\theta = \omega_n t \text{ and } x = r \cos \omega_n t \quad (8.1)$$

Differentiating: $\quad \dfrac{dx}{dt} = -\omega_n r \sin \omega_n t. \quad (8.2)$

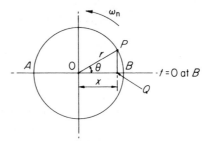

Fig. 8.4 Simple harmonic motion circle

Differentiating again: $\dfrac{d^2x}{dt^2} = -\omega_n^2 r \cos \omega t$.

Now $\dfrac{dx}{dt} = v$ the velocity of Q at any instant,

so that $\dfrac{d^2x}{dt^2} = \dfrac{dv}{dt} = a$, the acceleration of Q at any instant.

From 8.1: $r \cos \omega_n t = x$.

$$\therefore \dfrac{d^2x}{dt^2} = -\omega_n^2 x$$

so that $\dfrac{d^2x}{dt^2} + \omega_n^2 x = 0$ \hfill (8.3)

and $a = -\omega_n^2 x$ \hfill (8.4)

$$\therefore \omega_n = -\sqrt{\dfrac{a}{x}} \hfill (8.5)$$

i.e. $a \propto x$ the displacement from O. The minus sign in Equation 8.4 indicating that a increases in the opposite sense to x means that the acceleration is directed towards O. This conforms with the definition of SHM. Note that both the velocity and acceleration vary sinusoidally. Equation 8.3 should be recognised as the standard differential equation of simple harmonic motion or vibration.

ω_n is undamped natural frequency, hence the suffix n to distinguish it from other frequencies.

General relationships

Linear frequency (f_n) Hz $= \dfrac{\omega_n}{2\pi}$

From Equation 8.5 $\quad f_n = \dfrac{1}{2\pi}\sqrt{\dfrac{a}{x}}$

$$f_n = \dfrac{1}{2\pi}\sqrt{\dfrac{\text{acceleration}}{\text{displacement}}} \hfill (8.6)$$

Periodic time (τ) $= \dfrac{1}{f} = \dfrac{2\pi}{\omega_n}$

$$\tau = 2\pi \sqrt{\dfrac{\text{displacement}}{\text{acceleration}}} \qquad (8.7)$$

Equations 8.6 and 8.7 are general expressions which may be used once a motion has been proved to be simple harmonic. This is achieved by expressing the motion in the form of Equation 8.3, where x is a variable and ω_n is a constant, or by establishing an expression for the acceleration in the form of equation 8.4, whichever is the most convenient.

For torsional or angular vibration or motion these equations become:

$$\dfrac{d^2\theta}{dt^2} + \omega_n^2 \theta = 0$$

and $\quad \alpha = -\omega_n^2 \theta$

respectively where θ is an angular displacement in radians and α is an angular acceleration in rad/s².

Example 8.1

A component in a machine has a mass of 2 kg and oscillates with SHM. The amplitude of the oscillation is 0.25 m and the periodic time is 3 s. Determine:

(a) the acceleration at 0.2 m from either extremity;
(b) the velocity at the mid-point;
(c) the accelerating force at the point in (a);
(d) the maximum acclerating force.

Since the amplitude is 0.25 m, 0.2 m is 0.05 m from the mid-point.

∴ From Equation 8.4:

$$a = -\omega_n^2 x$$

$x = 0.05$ m
$\omega_n = 2\pi/\tau = 2\pi/3$ s

$$\therefore a = \left(\dfrac{2\pi}{3}\right)^2 \times 0.05 = 0.219 \text{ m/s}^2$$

Now $v = -\omega_n r \sin \omega_n t \quad$ at the mid-point.

$$\omega_n t = \dfrac{\pi}{2} \text{ rad}$$

$$\sin \omega_n t = \sin \dfrac{\pi}{2} = 1$$

$$\therefore v = -\omega_n r$$

$$= -\dfrac{2\pi}{3} \times 0.25 = 0.524 \text{ m/s}$$

From $p = ma$, $a = 0.219$ m/s² at 0.2 m from extremities.

$$\therefore p = 2 \times 0.219 = 0.438 \text{ N}.$$

Maximum acceleration from 8.4:

$$a_{max} = -\omega_n^2 r$$
$$= -\left(\frac{2\pi}{3}\right)^2 \times 0.25$$
$$= 1.1 \text{ m/s}^2$$
$$\therefore p_{max} = 2 \times 1.1 = 2.2 \text{ N}.$$

Example 8.2
A flat spring, arranged as a cantilever, supports an end load of mass m kg. With the load stationary, the deflection at the free end is y m. The mass is disturbed by a momentary force which deflects the free end a further x m from the equilibrium position.

(a) Prove that the subsequent vibration is simple vibration.

(b) If $m = 40$ g; $y = 3$ mm and $x = 2$ mm determine the frequency of the vibration assuming no damping.

(a) Static force exerted by the mass on the spring $= mg$ N
where g is the accleration due to gravity.

$$\text{Spring rate} = mg/y \text{ N/m}$$
$$\therefore \text{ Momentary disturbing force} = \frac{mg}{y} x \text{ N}.$$

This force will act on the mass but it will be resisted by the inertia force exerted by the mass.

$$\text{Inertia force} = ma \text{ or } m\frac{d^2x}{dt^2}$$

For equilibrium:

$$m\frac{d^2x}{dt^2} + \frac{mg}{y} x = 0$$
$$\therefore \frac{d^2x}{dt^2} + \frac{mg}{my} x = 0$$

This is the SHM form of Equation 8.3.

$$\therefore \omega_n^2 = g/y$$
$$\omega_n = \sqrt{g/y}.$$

Alternatively, equating inertia and disturbing forces:

$$ma = -\frac{mg}{y} x$$
$$a = -g/y x$$

Thus, $a \propto x$ so the motion is SHM.

(b) $\omega_n = \sqrt{g/y}$
$g = 9.81$ m/s²
$y = 3$ mm

$$\omega_n = \sqrt{\frac{9.81}{3 \times 10^{-3}}} = 57.18 \text{ rad/s}$$

$$f_n = \frac{\omega_n}{2\pi} = \frac{57.18}{2\pi} = 9.1 \text{ Hz.}$$

It should be noted that ω_n is independent of the initial momentary disturbance and depends only g and y.

8.4 Vibration of a shaft with one end fixed with the other carrying an inertia

The wheel, which has a moment of inertia I, is fixed to the lower end of the shaft as shown in Fig. 8.5.

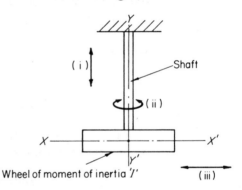

Fig. 8.5 Diagram of the wheel and shaft

Displacement of the wheel can occur, so that vibrations may result in 4 possible modes shown in Fig. 8.5:
 (i) linear vibrations in the vertical plane along the axis YY';
 (ii) torsional vibrations about the axis YY';
 (iii) & (iv) for small displacements, approximately linear vibrations on the axis XX' or ZZ'.
These will be dealt with separately. Of the four possibilities it is likely that those in (ii), (iii) and (iv) will be more significant in most practical cases.

(i) Longitudinal vibration

In Fig. 8.6:
  m = mass of wheel;
  l = length of shaft;
  E = modulus of elasticity of shaft material;
  A = cross-sectional area of shaft;
  x is the displacement caused by an initial disturbing force F.

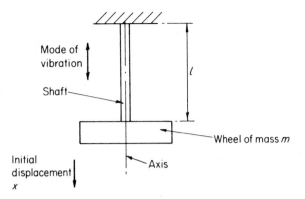

Fig. 8.6 Longitudinal vibration

In the tension formula $E = \dfrac{Wl}{Ax}$, in this case $W = F$

$$\therefore F = \dfrac{EAx}{l}.$$

This force will be acting on the mass to return it to the original position. The wheel, due to the inertia of its mass, will offer resistance to motion.

$$\text{Inertial force} = m\dfrac{d^2x}{dt^2}$$

$\therefore$ For equilibrium:

$$m\dfrac{d^2x}{dt^2} + \dfrac{EA}{l}x = 0$$

$$\dfrac{d^2x}{dt^2} + \dfrac{EA}{lm}x = 0$$

This is the SHM form of Equation 8.3.

$$\therefore \omega_n^2 = \dfrac{EA}{lm} \quad \text{and} \quad \omega_n = \sqrt{\dfrac{EA}{lm}} \text{ rad/s} \qquad (8.8)$$

$$f_n = \dfrac{1}{2\pi}\sqrt{\dfrac{EA}{lm}} \text{ Hz} \qquad (8.9)$$

It should be noted that $\dfrac{EA}{l} = \dfrac{W}{x} = $ longitudinal stiffness of the shaft

so that $\omega_n = \sqrt{\dfrac{\text{longitudinal stiffness of support member}}{\text{mass supported}}}$

(ii) Torsional vibration

In Fig. 8.7:
  $I = $ moment of intertia of the wheel about axis YY';
  $G = $ modulus of rigidity of the shaft material;

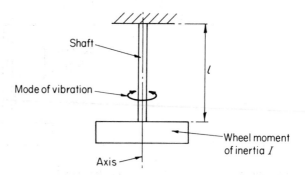

Fig. 8.7 Torsional vibration

J = polar 2nd moment of area of shaft;
l = length of shaft;
θ is the initial angular displacement caused by a disturbing torque T;

From the torsion formula: $\dfrac{T}{J} = \dfrac{G\theta}{l}$

$$\therefore T = \dfrac{GJ\theta}{l}$$

This torque will be acting on the wheel to return it to its original undisturbed position. The wheel, due to its moment of inertia, will offer resistance to motion.

$$\text{Resisting or inertia torque} = I\dfrac{d^2\theta}{dt^2}$$

∴ For equilibrium:
$$I\dfrac{d^2\theta}{dt^2} + \dfrac{GJ}{l}\theta = 0$$

and
$$\dfrac{d^2\theta}{dt^2} + \dfrac{GJ}{Il}\theta = 0$$

This is the SHM form of Equation 8.3.

$$\therefore \omega_n^2 = \dfrac{GJ}{Il}$$

and
$$\omega_n = \sqrt{\dfrac{GJ}{Il}} \text{ rad/s} \qquad (8.10)$$

$$f_n = \dfrac{1}{2\pi}\sqrt{\dfrac{GJ}{Il}} \text{ Hz} \qquad (8.11)$$

Again it should be noted that $\dfrac{GJ}{l} = \dfrac{T}{\theta}$ = torsional stiffness of the shaft.

$$\therefore \omega_n = \sqrt{\dfrac{\text{torsional stiffness of supporting member}}{\text{moment of inertia of supported mass}}}$$

(iii) Lateral vibration

In this case it is assumed that the length of the shaft is large compared with the amplitude of vibration so that the vibrations may be assumed to be linear.

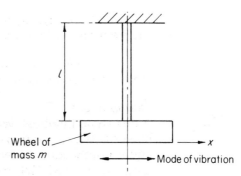

Fig. 8.8 Lateral vibration

m, E and I are defined in Section 8.4 (i).
I is the second moment of area of shaft section about a diameter (i.e. $\pi d^4/64$).

From Chapter 3 the deflection of a cantilever due to an end load W is $\dfrac{Wl^3}{3EI}$.

In this instance the deflection x is caused by an initial disturbing force F so that

$$x = \frac{Fl^3}{3EI} \quad \text{and} \quad F = \frac{3EIx}{l^3}.$$

This force acts on the mass to return it to its original position with the shaft vertical. The mass, due to its inertia, will offer resistance to motion.

$$\text{Inertia force} = \frac{m \mathrm{d}^2 x}{\mathrm{d}t^2}$$

$$\therefore \quad \frac{m \mathrm{d}^2 x}{\mathrm{d}t^2} + \frac{3EIx}{l^3} = 0$$

Divide by M:

$$\frac{\mathrm{d}^2 x}{\mathrm{d}t^2} + \frac{3EI}{ml^3} x = 0$$

This is the SHM form of Equation 8.3.

$$\therefore \quad \omega_n^2 = \frac{3EI}{ml^3}$$

$$\omega_n = \sqrt{\frac{3EI}{ml^3}} \text{ rad/s} \tag{8.12}$$

$$f_n = \frac{1}{2\pi} \sqrt{\frac{3EI}{ml^3}} \text{ Hz.} \tag{8.13}$$

The lateral or bending stiffness of the shaft is $\frac{E}{x} = \frac{3EI}{l^3}$,

so that $\omega_n = \sqrt{\frac{\text{lateral stiffness}}{\text{mass supported}}}$ rad/s.

From sections (i), (ii) and (iii), it is reasonable to conclude that the frequency of vibration in each mode depends upon

$$\sqrt{\frac{\text{stiffness of supporting member in that mode}}{\text{mass (for linear motion) or moment of inertia (for angular motion)}}} \text{ rad/s.}$$

Example 8.3
A disc-type flywheel, with a radius of gyration of 150 mm and a mass of 20 kg is fixed to the lower end of a vertical shaft, the upper end is rigidly constrained. The shaft is 50 mm diameter and 0.3 m long. If $G = 80$ GN/m² and $E = 207$ GN/m² determine the frequency of vibration in each of the possible modes.

Moment of inertia $(I) = mk^2 = 20 \times 0.15^2 = 0.45$ kg m².

Longitudinal stiffness of the shaft $= \frac{Ea}{l}$

$$= \frac{207 \times 10^9 \times \pi/4 \times 0.05}{0.3^2}$$

$= 1.3 \times 10^9$ N/m.

Lateral stiffness $= \frac{3EI}{l^3} = \frac{3 \times 207 \times 10^9 \times \pi/64 \times 0.05^4}{0.3^3}$

$= 7.05 \times 10^8$ N/m.

Torsional stiffness $= \frac{GJ}{l} = \frac{80 \times 10^9 \times \pi/32 \times 0.05^4}{0.3}$

$= 164 \times 10^3$ Nm/rad.

∴ Frequency of longitudinal vibration $= \frac{1}{2\pi} \sqrt{\frac{1.3 \times 10^9}{20}}$

$= 1283$ Hz.

Frequency of lateral vibration $= \frac{1}{2\pi} \sqrt{\frac{7.05 \times 10^8}{20}} = 94.5$ Hz.

Frequency of torsional vibration $= \frac{1}{2\pi} \sqrt{\frac{164 \times 10^3}{0.45}} = 96$ Hz.

Example 8.4
A vertical steel wire 0.9 m long and 2.5 mm diameter is fixed at its upper end and carries a short solid cast iron cylinder at its lower end. The cylinder has a mass of 6 kg and a diameter of 0.2 m. If $G = 70$ GN/m² and the cylinder is rotated 5° about the vertical axis and released determine:

(a) the frequency of the subsequent oscillations;
(b) the maximum angular velocity of the cylinder.

For angular motion:

$$f_n = \frac{1}{2\pi}\sqrt{\frac{\text{torsional stiffness of the wire}}{\text{moment of inertia of the cylinder}}}$$

Torsional stiffness $\left(\frac{T}{\theta}\right) = \frac{GJ}{l} = \frac{70 \times 10^9 \times \pi/32 \times 0.0025^4}{0.9}$

$$= 0.298 \text{ Nm/rad}$$

Moment of inertia $I = mk^2 = 6 \times \frac{0.1^2}{2}$

$$= 0.03 \text{ kg m}^2$$

NB $\left(k = \frac{r}{\sqrt{2}}; k^2 = \frac{r^2}{2}\right)$

$$\therefore f_n = \frac{1}{2\pi}\sqrt{\frac{0.298}{0.03}} = 0.5 \text{ Hz}$$

From Equation 8.2 $v = \omega_n r \sin \omega_n t$
when $\sin \omega_n t = 1$ and r is the amplitude of vibration:

$$V_{max} = \omega_n r$$

$\omega_n = 3.15$ rad/s; $r = 5° = 0.0873$ rad.

$$\therefore V_{max} = 3.15 \times 0.0873 = 0.275 \text{ rad/s}$$

8.5 Transverse vibration of a beam with a single concentrated load

Components which act as simply supported beams carrying static loads are not uncommon in machines. An additional momentary disturbing force may be sufficient to set them into vibration. In this section the case of the single concentrated load is dealt with.

Referring to Fig. 8.9 and to Table 3.1, in Chapter 3, the deflection under the load W is given by:

$$y = \frac{Wa^2b^2}{3EI(a+b)} = \frac{Wa^2b^2}{3EIl}$$

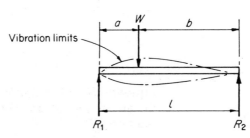

Fig. 8.9 Diagram of the beam showing vibration mode

∴ The stiffness of the beam under the load $= \dfrac{W}{y} = \dfrac{3EII}{a^2b^2}$.

The force required to cause an additional momentary deflection of $y_1 = \dfrac{3EII}{a^2b^2} y_1$. This force will act on the load mass to return it to the position of static deflection. Due to its inertia the load mass will offer resistance to motion.

$$\text{Inertia force} = m\dfrac{d^2y}{dt^2} \text{ where } m = W/g$$

∴ Equating the two forces gives

$$m\dfrac{d^2y}{dt^2} = -\dfrac{3EII}{a^2b^2}y_1$$

or $\quad m\dfrac{d^2y}{dt^2} + \dfrac{3EII}{a^2b^2}y_1 = 0$

and $\quad \dfrac{d^2y}{dt^2} + \dfrac{3EII}{ma^2b^2}y_1 = 0$

this is the SHM form of Equation 8.3.

$$\therefore \omega_n = \sqrt{\dfrac{3EII}{ma^2b^2}} \text{ rad/s} \tag{8.14}$$

$$f_n = \dfrac{1}{2\pi}\sqrt{\dfrac{3EII}{ma^2b^2}} \text{ Hz.} \tag{8.15}$$

Once again $\quad \omega_n = \sqrt{\dfrac{\text{deflection stiffness}}{\text{mass of load}}}$

Since $y = \dfrac{Wa^2b^2}{3EII}$, in Equations 8.14 and 8.15 $\dfrac{3EII}{ma^2b^2}$ can be written as $\dfrac{g}{y}$

$$\therefore \omega_n = \sqrt{g/y} \tag{8.16}$$

and $\quad f_n = \dfrac{1}{2\pi}\sqrt{g/y} \tag{8.17}$

In Equation 8.17, since $\dfrac{\sqrt{g}}{2\pi} = 0.498$ or 0.5 approximately,

$$f_n = \dfrac{0.5}{\sqrt{y}} \tag{8.18}$$

Equation 8.18 is a particularly useful form as it facilitates the use of an empirical formula for general loading dealt with in Section 8.7.

Example 8.5
A shaft which is 30 mm diameter is supported in spherical bearings over a span of 1 m. A wheel of mass 180 kg is placed:
  (a) 0.225 m from one bearing;
  (b) at the centre.
If $E = 200$ GN/m^2 and $g = 9.81$ m/s^2 determine the frequency of transverse vibration in each case.

(a) y = deflection under load
$$= \frac{Wa^2b^2}{3EIl}$$

$W = mg = 180 \times 9.81 = 1766$ N; $a = 0.225$ m; $b = 0.775$ m;

$$I = \frac{\pi}{64} \times 0.03^4 = 3.98 \times 10^{-8} \text{ m}^4; l = 1 \text{ m}.$$

$$\therefore y = \frac{1766 \times 0.225^2 \times 0.775^2}{3 \times 200 \times 10^9 \times 3.98 \times 10^{-8} \times 1} = 2.25 \times 10^{-3} \text{ m}$$

Using Equation 8.18:

$$f_n = \frac{0.5}{\sqrt{2.25 \times 10^{-3}}} = 10.5 \text{ Hz}.$$

(b) $$y = \frac{Wl^3}{48EI} \text{ under the centre load}$$

$$= \frac{1766 \times 1^3}{48 \times 200 \times 10^9 \times 3.98 \times 10^{-8}}$$

$$= 4.62 \times 10^{-3} \text{ m}.$$

$$f_n = \frac{0.5}{\sqrt{4.62 \times 10^{-3}}} = 7.35 \text{ Hz}.$$

8.6 Transverse vibration of a beam carrying a distributed load

In Fig. 8.10 w = load/m and l = length of beam. As the load is distributed, its movement during vibration varies from zero, at the

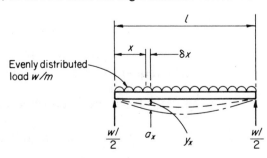

Fig. 8.10 Diagram showing the transverse vibration of a beam carrying a distributed load

supports, to a maximum at the centre. Consequently the methods used in the previous cases are not valid. The method which is to be used recognises that, at the mid-point of vibration, the load has maximum velocity and maximum kinetic energy, whereas at the extremities of vibration it is momentarily stationary, and all its kinetic energy has been converted into strain energy in the beam material. Assuming no losses:

$$\begin{matrix}\text{kinetic energy of load} \\ \text{at the mid-point of vibration}\end{matrix} = \begin{matrix}\text{strain energy of beam} \\ \text{at the extremes of vibration.}\end{matrix}$$

Expressions are derived for these energies for a small element of the load, so that the amplitude of the vibration may be considered as constant. By subsequent integration, an expression applicable to the whole beam is derived.

Consider an element of the load covering a length δx of the beam and distance x from one end, Fig. 8.10. Let the static deflection at this point of the beam due to the load be y_x, and the additional maximum dynamic deflection due to the vibration be a_x.

If it is assumed that the static and dynamic deflection curves are of similar profiles, which is not unreasonable, then

$$a_x/y_x = \text{a constant } (k) \text{ thus } a_x = ky_x.$$

$$\text{Stiffness of beam at this section} = \frac{w\delta x}{y_x}$$

$$\therefore \text{ Force to cause a dynamic deflection} = \frac{w\delta x}{y_x}$$

$$\text{Maximum strain energy of element} = \left(\frac{w\delta x}{y_x}ax\right)\frac{ax}{2}$$

$$= \frac{w\delta x}{2y_x}ax^2 = \delta U_{max}$$

Substituting: $a_x = ky_x$

$$\therefore \delta U_{max} = \frac{wk^2}{2}y_x\delta x$$

$$\text{Total strain energy of the beam} \equiv \frac{wk^2}{2}\int_0^l y_x dx.$$

The mass of the element of the load $= \frac{w\delta x}{g}$, and its velocity at the mid-point of vibration is $\omega_n a_x$ where ω_n is the angular frequency of natural vibration and a_x is the amplitude of that vibration.

$\therefore$ From the expression for kinetic energy, $KE = \frac{1}{2}mv^2$

$$\text{KE of the element} = \frac{w\delta x}{2g}ax^2\omega_n^2.$$

Substituting ky_x for a_x: $= \dfrac{wk^2\omega_n^2}{2g} y_x^2 \delta x$

Total KE for whole load $= \dfrac{wk^2\omega_n^2}{2g} \int_0^l y_x^2 \, dx$

Equating the kinetic energy to the strain energy:

$$\dfrac{wk^2\omega_n^2}{2g} \int_0^l y_x^2 \, dx = \dfrac{wk^2}{2} \int_0^l y_x \, dx$$

$$\therefore \omega_n = \sqrt{\dfrac{g \int_0^l y_x \, dx}{\int_0^l y_x^2 \, dx}} \quad (8.19)$$

$$f_n = \dfrac{1}{2\pi} \sqrt{\dfrac{g \int_0^l y_x \, dx}{\int_0^l y_x^2 \, dx}} \quad (8.20)$$

From Table 3.1 in Chapter 3, the expression for the deflection of a simply supported beam with a distributed load is:

$$y_x = \dfrac{1}{EI}\left(\dfrac{wx^4}{24} - \dfrac{wlx^3}{12} + \dfrac{wl^3 x}{24}\right)$$

which may be simplified to:

$$y_x = \dfrac{w}{24EI}(x^4 - 2lx^3 + l^3 x) \quad (8.21)$$

Squaring 8.21 gives:

$$y_x^2 = \dfrac{w^2}{576 E^2 I^2}(x^8 - 4lx^7 + 2l^3 x^5 - 4l^2 x^6 - 4l^4 x^4 + l^6 x^2) \quad (8.22)$$

Integrating 8.21 between limits gives: $\dfrac{wl^5}{120EI}$

Integrating 8.22 between limits gives: $\dfrac{w^2 l^9}{11\,700 EI}$

Substituting these values in equation 8.19:

$$\omega_n = \sqrt{\dfrac{g\dfrac{wl^5}{120EI}}{\dfrac{w^2 l^9}{11\,700 EI}}} = \sqrt{\dfrac{97.5\, EIg}{wl^4}}$$

which is a general expression for the frequency of vibration. However since the maximum deflection due to the load is $y_{max} = \dfrac{5}{384}\dfrac{wl^4}{EI}$, comparing this with $\dfrac{wl^4}{97.5\, EI}$ in the general expression for ω_n above it is apparent that:

$$\dfrac{wl^4}{97.5\, EI} = 0.787\, y_{max} \quad \text{or} \quad \dfrac{y_{max}}{1.27}.$$

It is thus possible to simplify the original general expression to:

$$\omega_n = \sqrt{\frac{g}{y_{max}/1.27}} \qquad (8.23)$$

$$f_n = \frac{1}{2\pi}\sqrt{\frac{g}{y_{max}/1.27}}. \qquad (8.24)$$

Example 8.7
A simply supported beam 0.75 m long is made of steel of density 7 mg/m³. The beam is circular in section 50 mm diameter. Assuming $E = 200$ GN/m² and $g = 9.81$ m/s² determine the frequency of transverse vibration if the beam is disturbed from the static deflection position.

$$\omega_n = \sqrt{\frac{g}{y_{max}/1.27}}$$

$$y_{max} = \frac{5}{384}\frac{wl^4}{EI}$$

$W = \frac{\pi}{4} \times 0.05^2 \times 1 \times 7 \times 10^3 \times 9.81$ N/m

$= 135$ N/m.

$l = 0.75$ m.

$E = 200 \times 10^9$ N/m².

$I = \frac{\pi}{64} \times 0.05^4 = 3.07 \times 10^{-7}$ m⁴.

$$y_{max} = \frac{5}{384} \times \frac{135 \times 0.75^4}{200 \times 10^9 \times 3.07 \times 10^{-7}} = 9.06 \times 10^{-6} \text{ m}.$$

From 8.23:

$$\omega_n = \sqrt{\frac{9.81}{\frac{9.06 \times 10^{-6}}{1.27}}} = 1172 \text{ rad/s}.$$

From 8.24:

$$f_n = \frac{1}{2\pi} \times 1172 = 187 \text{ Hz}.$$

8.7 Dunkerley's Formula

To deal with a beam with a combination of loads would become mathematically very involved but for an empirical formula attributed to Dunkerley. This is based on the calculation of the frequency of vibration for each individual load, considered separately. If these individual frequencies are ω_1; ω_2 etc for concentrated loads and ω_s for the distributed load due to the weight of the beam then:

$$\frac{1}{\omega_n^2} = \frac{1}{\omega_1^2} + \frac{1}{\omega_2^2} \ldots + \frac{1}{\omega_s^2}$$

where ω_n is the overall frequency of vibration.

Dunkerley applied

Referring to Fig 8.11 for the simply supported beam or shaft let:
ω_1 = frequency of vibration due to load W_1;
ω_2 = frequency of vibration due to load W_2;
ω_s = frequency of vibration due to the distributed load W.

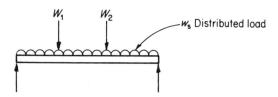

Fig. 8.11 Diagram of the simply supported beam or shaft

The corresponding deflections are y_1; y_2 and y_{max} respectively. Then, from 8.16:

$$\frac{1}{\omega_1^2} = \frac{y_1}{g}; \quad \frac{1}{\omega_2^2} = \frac{y_2}{g}.$$

From 8.23:

$$\frac{1}{\omega_s^2} = \frac{y_{max}/1.27}{g}.$$

$$\therefore \frac{1}{\omega_n^2} = \frac{y_1}{g} + \frac{y_2}{g} + \frac{y_{max}/1.27}{g}$$

$$= \frac{1}{g}(y_1 + y_2 + y_{max}/1.27)$$

$$\omega_n = \sqrt{\frac{g}{y_1 + y_2 + y_{max}/1.27}}$$

$$f_n = \frac{\sqrt{g}}{2\pi}\sqrt{\frac{1}{y_1 + y_2 + y_{max}/1.27}}$$

$$= \frac{0.5}{\sqrt{y_1 + y_2 + y_{max}/1.27}} \text{ Hz.} \qquad (8.25)$$

Example 8.8
A steel shaft which is 75 mm diameter is supported in self-aligning bearings which are 1.6 m apart. The shaft carries 3 gears, each of mass 90 kg, which are equally spaced between the bearings. Ignoring the weight of the shaft determine the frequency of transverse vibration. $E = 200 \text{ GN/m}^2$.

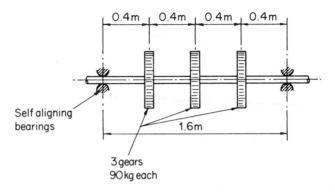

Fig. 8.12 The steel shaft supported in self aligning bearings and carrying three gears

Referring to Fig. 8.12:
$W_1 = W_2 = W_3 = 90 \times 9.81 = 883$ N.

Static deflection due to W_1 = Static deflection due to W_3

i.e. $$y_1 = y_3 = \frac{Wa^2b^2}{3EIl}$$

where $W = 883$ N; $a = 0.4$ m; $b = 1.2$ m; $E = 200 \times 10^9$ N/m²;
$I = \frac{\pi}{64} \times 0.075^4 = 1.55 \times 10^{-6}$ m⁴; $l = 1.6$ m.

$$\therefore y_1 = y_3 = \frac{883 \times 0.4^2 \times 1.2^2}{3 \times 200 \times 10^9 \times 1.55 \times 10^{-6} \times 1.6}$$

$$= 1.37 \times 10^{-4} \text{ m}.$$

Static deflection due to $W_2 = y_2 = \dfrac{Wl^3}{48EI}$

$$= \frac{883 \times 1.6^3}{48 \times 200 \times 10^9 \times 1.55 \times 10^{-6}}$$

$$= 2.43 \times 10^{-4} \text{ m}.$$

Using Equation 8.25

$$f_n = \frac{0.5}{\sqrt{y_1 + y_2 + y_3}}$$

$$= \frac{0.5}{\sqrt{1.37 \times 10^{-4} + 2.43 \times 10^{-4} + 1.37 \times 10^{-4}}}$$

$$= 22 \text{ Hz}$$

Example 8.9
If the shaft in Example 8.8 had a density of 7 Mg/m³, determine the frequency of vibration taking the weight of the shaft into account. Assume $g = 9.81$ m/s².

Mass/unit length $= \dfrac{\pi}{4} \times 0.075^2 \times 1 \times 7 \times 10^3 = 30.9$ kg/m

Weight/unit length $= 30.9 \times 9.81 = 303$ N/m

Deflection due to the distributed load $= \dfrac{5\,wl^4}{384\,EI}$

$$= \dfrac{5 \times 303 \times 1.6^4}{384 \times 200 \times 10^9 \times 1.55 \times 10^{-6}}\,\text{m}$$

$y_{max} = 8.34 \times 10^{-5}\,\text{m}$

$\therefore\ y_{max}/1.27 = \dfrac{8.34}{1.27} \times 10^{-5} = 6.57 \times 10^{-5}\,\text{m}.$

$\therefore\ f_n =$

$$\dfrac{0.5}{\sqrt{1.37 \times 10^{-4} + 2.43 \times 10^{-4} + 1.37 \times 10^{-4} + 6.57 \times 10^{-5}}}$$

$= 20.7\,\text{Hz}.$

8.8 Whirling of shafts

A shaft is said to *whirl* when it rotates in a *bowed* condition between the bearings. Under certain conditions if the speed of rotation of a shaft is gradually increased from zero, it will be seen that approaching a particular speed the shaft will begin to bow until the bowing is a maximum at that speed. If the speed is further increased the *bow* disappears, only to reappear in slightly different forms at higher speeds. The speeds at which the bowed condition is a maximum are known as *critical* or *whirling* speeds and they represent a dangerous condition which needs to be avoided. Since even with careful balancing, whirling cannot be avoided the critical speeds need to be identified. Any equipment employing lengths of shafting and high speeds is particularly prone to the problem. Turbine rotors and shafts, marine propellor shafts are two examples.

The reasons for whirling originate in the need to provide centripetal acceleration for a mass to rotate in a circular path. This will arise when the centre of gravity of a rotating mass does not coincide with the axis of rotation. For a shaft this may be due to:
(i) imperfect machining;
(ii) deflection due to the weight of the shaft;
(iii) deflection due to a mass attached to the shaft etc.
Whatever the cause, it is the strength of the shaft material which must provide the centripetal force to give the required acceleration. At the critical, or whirling, speeds the elastic stiffness of the shaft is just able to provide the necessary centripetal force and a condition of unstable equilibrium is reached where failure may eventually result.

At this stage the study of the whirling phenomena will be confined to the consideration of a light shaft with a single eccentric rotor attached. However the results will suggest a simple way of identifying whirling speeds in more complicated cases.

(i) Simply supported shaft with single eccentric load

Referring to Fig. 8.13:

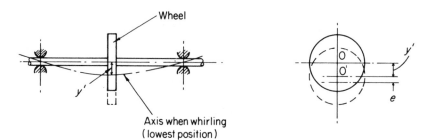

Fig. 8.13 The simply supported shaft with a single eccentric load

m = mass of rotor;
W = weight of rotor = mg;
e = eccentricity of the centre of gravity; of the rotor;
y = the static deflection of the shaft due to the weight of the disc;
y' = the dynamic deflection due to the eccentricity.

The shaft weight is ignored.

The deflection stiffness of the shaft = $\dfrac{W}{y}$ at the rotor position. At any angular velocity of ω rad/s:

$$\text{centripetal force required} = m\omega^2(y' + e)$$

$$\text{elastic force provided by shaft material} = \dfrac{W}{y} y'$$

Equating the forces:

$$\dfrac{W}{y} y' = m\omega^2 (y' + e).$$

Put $W = mg$ and divide by m:

$$g \dfrac{y'}{y} = \omega^2 y' + \omega^2 e.$$

But from Equation 8.16 in Section 8.5, $\dfrac{g}{y} = \omega_n^2$ where ω_n is the undamped frequency of natural vibration:

$$\therefore \omega_n^2 y' = \omega^2 y' + \omega^2 e$$

$$y' = \dfrac{\omega^2 e}{(\omega_n^2 - \omega^2)}$$

When $\omega \to \omega_n$ then $(\omega_n^2 - \omega^2) \to 0$ and y' becomes infinite; this is the critical condition. The first critical whirling speed coincides, therefore, with the natural frequency of lateral vibration of the shaft carrying the rotor.

This principle is confirmed by more mathematically involved proofs, with other types of loading the first whirling speeds can be calculated using the Dunkerley formula developed in Section 8.7. Subsequent whirling speeds are related to the first whirling speed which is thus very important.

Example 8.10
A pulley with a mass of 70 kg is fixed to a shaft 0.5 m long and 25 mm diameter, midway between the bearings. Ignoring the weight of the shaft, determine the first critical whirling speed. Assume $E = 200$ GN/m² and $g = 9.81$ m/s².

From Equation 8.16 in section 8.5:

$$\omega_n = \sqrt{\frac{g}{y}};$$

$$y = \frac{Wl^3}{48EI}.$$

$W = mg = 70 \times 9.81 = 687$ N; $l = 0.5$ m; $E = 200 \times 10^9$ N/m²;

$I = \frac{\pi}{64} \times 0.025^4 = 1.92 \times 10^{-8}$ m⁴.

$$\therefore y = \frac{687 \times 0.5^3}{48 \times 200 \times 10^9 \times 1.92 \times 10^{-8}} = 4.66 \times 10^{-4} \text{ m}.$$

$$\omega_n = \sqrt{\frac{g}{y}}$$

$$= \sqrt{\frac{9.81}{4.66 \times 10^{-4}}} = 145 \text{ rad/s or } 1386 \text{ rev/min}.$$

8.9 Summary

(i) *Simple vibration*

$$\frac{d^2x}{dt^2} + \omega_n^2 x = 0$$

$$\omega_n = 2\pi f = \sqrt{\frac{\text{acceleration}}{\text{displacement}}}$$

$$\tau = \frac{1}{f}$$

(ii) *Shaft with one end fixed and rotor at the other end*
Longitudinal vibration:

$$\omega_n = \sqrt{\frac{\text{longitudinal stiffness}}{\text{mass supported}}} \text{ rad/s}.$$

Lateral vibration:
$$\omega_n = \sqrt{\frac{\text{lateral stiffness}}{\text{mass supported}}} \text{ rad/s.}$$

Torsional vibration:
$$\omega_n = \sqrt{\frac{\text{torsional stiffness}}{\text{inertia of moving mass}}} \text{ rad/s.}$$

(iii) *Lateral vibration of simply supported shaft or beam*

Single mass
$$\omega_n = \sqrt{\frac{g}{y}} \text{ rad/s} \quad f_n = \frac{0.5}{\sqrt{y}} \text{ Hz}$$

Shaft mass alone
$$\omega_n = \sqrt{\frac{g}{y_{max}/1.27}} \text{ rad/s} \quad f_n = \frac{0.5}{\sqrt{y_{max}/1.27}} \text{ Hz}$$

Dunkerley
$$f_n = \frac{0.5}{\sqrt{y_1 + y_2 + \ldots y_{max}/1.27}} \text{ Hz.}$$

(iv) *First critical whirling speed (rad/s)* = undamped frequency of natural vibration (rad/s)

Exercises 8

Questions 1 to 5 refer to the theory and should be attempted initially without further reference to the text.

1 In connection with vibrations and the concept of 'degrees of freedom' indicate which of the following statements are correct.
   (a) A body has a maximum of 3 degrees of freedom.
   (b) Vibrations may consist of a combination of movements in all three planes.
   (c) Ignoring mechanical strain, a nut on a thread has two degrees of freedom.
   (d) Ignoring mechanical strain and end float, a gearwheel on a shaft has only one degree of freedom.
   (e) A wheel running in a straight groove has only one degree of freedom.

2 Indicate the correct statements in the following.
   (a) Simple vibration is synonymous with simple harmonic motion.
   (b) For simple vibration theory, damping is completely disregarded.

(c) The periodic time for a simple vibration depends only on displacement and acceleration.
(d) The velocity of motion is a maximum when the acceleration is zero.
(e) Displacement, velocity and acceleration are sinusoidal functions for both linear and angular motion.

3 In connection with the various modes of vibration indicate which of the following statements are correct.
 (a) For all three modes the mass inversely affects the natural frequency.
 (b) For all three modes increasing the mode stiffness will increase the frequency.
 (c) The frequency of vibration is independent of the original disturbing force.
 (d) For the torsional mode, the stiffness
$$= \frac{\text{torque}}{\text{angular displacement}}$$
 (e) For the longitudinal mode, the stiffness
$$= \frac{\text{modulus of elasticity} \times \text{cross sectional area}}{\text{length}}$$

4 For the transverse vibration of a beam, certain of the following statements are correct. Indicate the correct statements.
 (a) The vibration depends upon the nature of the supports.
 (b) The vibration depends upon the deflection stiffness of the beam.
 (c) For a single concentrated load the frequency of vibration is independent of its mass.
 (d) Using the Dunkerley method, the vibration for each load is calculated separately.
 (e) Dunkerley uses an empirical formula.

5 In connection with the whirling of shafts, indicate which of the following statements are correct.
 (a) Whirling is said to occur when a shaft rotates in a 'bowed' condition between bearings.
 (b) The whirling speed is related directly to the lateral frequency of vibration.
 (c) Whirling may be due to imperfect machining.
 (d) Whirling results in a condition of unstable equilibrium at critical speeds.
 (e) Whirling speeds depend upon the nature of the bearings.

continued

6 A cantilevered spring, for a timing mechanism, is to be arranged so that it vibrates with a periodic time of 1/5 s. The spring is to carry an end load of 300 g and is to be made from spring steel strip 5 mm × 1 mm. If $g = 9.81$ m/s² and $E = 200$ GN/m² determine the length of spring which must protrude.

7 A 3 mm diameter steel wire is fixed at its top end and carries at its lower end a short solid cylinder with its axis vertical. The cylinder is 0.15 m diameter and has a mass of 5 kg. If the effective length of the wire is 0.8 m and $G = 70$ GN/m² determine:
   (i) the frequency of oscillation about the vertical axis;
   (ii) the maximum angular velocity if the initial displacement is 10°.

8 A steel disc which is 0.25 m in outside diameter and has a mass of 16 kg is suspended vertically from its centre by a 3 mm diameter rod which is 1.5 m long. The upper end of the rod is rigidly fixed. If the disc is set in torsional oscillation, ignoring the inertia of the rod determine:
   (i) the frequency of torsional oscillation;
   (ii) the maximum angular velocity and acceleration if the amplitude of the oscillation is 82°.
$G = 80$ GN/m²

9 A shaft which is 30 mm diameter and 0.4 m long is made of a material which has a modulus of elasticity of 150 GN/m² and a modulus of rigidity of 55 GN/m². The shaft is fixed vertically and carries a gearwheel at its lower end. The wheel has a mass of 15 kg and a radius of gyration of 120 mm. Determine:
   (i) the stiffness of the shaft in each mode of vibration;
   (ii) the frequency of vibration in each mode.

10 (a) A torsional pendulum consists of a shaft 5 mm diameter and 0.75 m long with a disc fixed to its lower end. The periodic time of oscillation is 0.5 s. If $G = 80$ GN/m², determine the moment of inertia of the disc.
   (b) A gearwheel of unknown moment of inertia but of mass 20 kg is then fixed symmetrically to the disc. The system then oscillates with a frequency of 1.35 Hz. Determine:
     (i) the moment of inertia of the wheel;
     (ii) the radius of gyration of the wheel.

11 A simply supported beam has a second moment of area of 20 800 cm⁴. The supports are 6 m apart and a concentrated load of 30 kN is placed 2.4 m from one end. Determine the frequency of transverse vibration if the beam is de-

flected from its static position. $E = 200$ GN/m² and $g = 9.81$ m/s².

12 A simply supported beam with a span of 7.5 m carries a concentrated load at its centre. The I value of the beam section is 99 900 cm⁴, $E = 200$ GN/m², $g = 9.81$ m/s². The frequency of transverse vibration is 7.5 Hz. Determine:
   (i) the deflection under the load;
   (ii) the magnitude of the load.

13 A shaft is supported in spherical bearings 0.5 m apart. The shaft is 15 mm diameter and has a modulus of elasticity of 200 GN/m². A gearwheel of weight 50 N is splined to the shaft and can move from a position at the shaft centre to 0.2 m from one bearing. Determine the range of frequency of transverse vibration to which the shaft may be subjected. Take $g = 9.81$ m/s².

14 If the shaft in question 13 was made of steel of density 7 Mg/m³ determine the frequency of transverse vibration due only to the weight of the shaft.

15 For the shaft in questions 13 and 14 determine the resultant frequency of transverse vibration if the wheel is:
   (i) at the centre;
   (ii) 0.2 m from one bearing.

16 A beam has a weight equivalent to 2 kN/m and a second moment of area of 41 600 cm⁴. It is simply supported at its ends over a span of 3 m and carries concentrated loads of 20 kN and 40 kN at distances of 0.75 m and 1.5 m respectively from the left-hand end. If $E = 200$ GN/m² determine the frequency of transverse vibration.

17 A 15 mm diameter shaft carries a gearwheel of mass 15 kg midway between the bearings which are spherical and 0.4 m apart. The shaft has a weight of 12 N/m. If $E = 200$ GN/m² and $g = 9.81$ m/s², determine the first critical whirling speed
   (i) ignoring the weight of the shaft;
   (ii) taking the weight of the shaft into consideration.

18 A shaft 12.5 mm diameter is supported in spherically seated bearings 0.3 m apart. It carries a disc of weight 130 N at a point 0.175 m from one bearing. If $E = 200$ GN/m² determine the first critical whirling speed. Ignore the weight of the shaft and assume that $g = 9.81$ m/s².

19 A shaft 25 mm diameter has bearings which act as simple supports and are 0.45 m apart. The shaft carries a pulley of weight 100 N 0.25 m from one bearing. If $E = 200$ GN/m² determine the first critical whirling speed.

9 Balancing

9.1 Introduction

In the previous chapter we considered the problem of vibration and how its effects could be reduced. In some cases, by appreciating the forces involved, it is possible to eliminate the actual causes of the vibration and hence, the vibration itself. In this chapter we will be studying one such case.

When a mass is to be set into angular, as opposed to linear, motion it must be subjected to a centripetal acceleration. If a is the acceleration, ω the angular velocity and r the radius of rotation then:

$$a = \omega^2 r$$

The acceleration acts radially towards the centre of rotation and the force associated with it, the centripetal force (**CF**) is given by:

$$\mathbf{CF} = m\omega^2 r$$

where m is the rotating mass. The reaction to the centripetal force is a centrifugal force acting radially outwards of the same magnitude.

Some common examples of rotating masses are gearwheels, armatures, turbine runners, grinding wheels and crankshafts, to mention only a few. These components or assemblies are supported in bearings which are designed to carry the static weight. Normally, any centripetal force is provided by the mechanical strength of the component or assembly itself. If, due to design or manufacturing shortcomings, defects arising in service etc., the centre of gravity of the rotating mass does not coincide with the axis of rotation then the bearings must provide some of the centripetal force and the corresponding centrifugal reaction. This will be a cyclical effect according to the angular position of the rotating mass and it will result in vibration. To avoid this it is necessary to restore the centre of gravity to a position coinciding with the axis of rotation. This is called balancing.

In earlier studies the reader will have learnt how to deal with masses all rotating in the same plane, ie., coplanar, but some quick revision may be helpful.

9.2 Balancing coplanar masses

The basic procedure is to identify the out of balance force and neutralise it by an equal and opposite force. Fig. 9.1(a) shows three coplanar masses, m_1, m_2 and m_3 fixed in the angular positions shown at radii from the axis of rotation r_1, r_2 and r_3 respectively. If the angular velocity is ω rad/s then the corresponding centrifugal forces are $m_1\omega^2 r_1$, $m_2\omega^2 r_2$, $m_3\omega^2 r_3$. Since ω^2 is common only the mr products, or mass moments, are considered. The vector sum of the mass moments is determined for one angular position of the masses see Fig. 9.1(b).

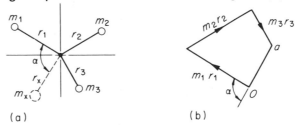

Fig. 9.1 (a) Space diagram
(b) Mass moment diagram

oa = vector sum or resultant in a direction $o \rightarrow a$.
∴ Out of balance centrifugal force = $oa \times \omega^2$
Balancing mass moment or equilibrant = oa in a direction $a \rightarrow o$.
If r_x is the radius of the balancing mass then the balancing mass

$m_x = \dfrac{oa}{r_x}$ and its angular position as in Fig. 9.1(a).

Example 9.1
Three coplanar masses of 5, 10 and 4 kg are fixed at radii of 0.15, 0.2 and 0.1 m respectively from the axis of the rotating shaft which carries them. When the 5 kg mass is vertically above the shaft axis the 10 kg mass is 90° clockwise from it and the 4 kg mass 135° clockwise. Determine:
(a) the out of balance force at 200 rev/min;
(b) the mass to be placed at a radius of 0.2 m to balance the system;
(c) the angular position of the mass in (ii) relative to the 5 kg mass.

Using a tabular method, which will be helpful later as well, the mass moments can be displayed as in Table 9.1.

Table 9.1.

mass (kg)	radius (m)	mass moment (kg m)
5	0.15	0.75
10	0.2	2.0
4	0.1	0.4

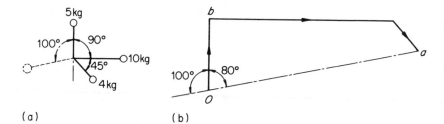

Fig. 9.2 (a) Space diagram
(b) Mass moment diagram

The mass moment diagram is drawn to scale in Fig. 9.2(b). In simple cases it can be solved algebraically.
From Fig. 9.2(b): $oa = 2.3$ kg m and $aob = 80°$

(a) Out of balance force $= oa \times \omega^2 = 2.3 \times \left(200 \times \dfrac{2\pi}{60}\right)^2$
$= 1009$ N.

(b) Balancing mass at 0.2 m radius $= \dfrac{2.3}{0.2} = 11.5$ kg.

(c) Angular position is $180° - 80° = 100°$ anticlockwise from the 5 kg mass as shown dotted in Fig. 9.2(a).

When coplanar systems are balanced in this way they will be balanced both dynamically and statically. This means that when the system is at rest it will remain in any angular position in which it is placed. This fact sometimes provides a quick way of achieving balance using movable balance weights.

9.3 Non-coplanar rotating masses

Fig. 9.3(a) and (b) shows two views of a system similar to that considered in Fig. 9.1 except that the masses are no longer coplanar. The planes of rotation of m_2 and m_3 are at distances of l_2 and l_3 respectively from that of m_1. As explained previously, the centrifugal forces will be out of balance but as they do not act in the same plane they will also produce a moment or couple. The couple may be calculated from any suitable reference plane.

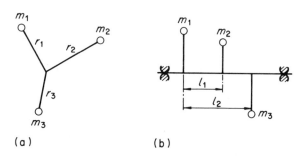

Fig. 9.3 (a) Space diagram
(b) Mass moment diagram

If we select the plane of m_1 as such a reference then the couples would be:
(i) due to m_1: zero;
(ii) due to m_2: (centrifugal force) $\times l_2$ or $(m_2\omega^2 r_2)l_2$;
(iii) due to m_3: $(m_3\omega^2 r_3)l_3$;
where ω is the angular velocity of the system.

If we select the plane of m_3 as a reference, the couples would be:
(i) due to m_1: $(m_1\omega^2 r_1)l_3$;
(ii) due to m_2: $(m_2\omega^2 r_2)(l_3 - l_2)$;
(iii) due to m_3: zero.

To find the resultant couple the individual couples must be added vectorially since the angular positions of the masses must be taken into account as well as their linear positions. Any such resultant will indicate an out of balance condition which will impose a cyclical force on the bearings and must therefore be eliminated. The 'couple' effect is only apparent when the system is rotating so that static balance of a *non-coplanar* system does not ensure dynamic balance.

Example 9.2
A shaft which is carried in bearings 0.5 m apart, has two wheels mounted on it which are out of balance. Wheel A has a mass of 40 kg and an eccentricity of its centre of gravity of 20 mm; wheel B has a mass of 60 kg with an eccentricity of 10 mm. Wheel A is 0.1 m and B is 0.35 m from the left-hand bearing. When the centre of gravity of wheel A is vertically above the shaft axis that of B is 120° clockwise when viewed from the right-hand end. Determine the dynamic reaction of each bearing when the shaft is rotating at 1200 rev/min.

Fig. 9.4(a) and (b) shows two views of the system. As there are two unknown forces, ie the reaction at each bearing, the plane of reference is taken through one bearing only. From the couple diagram, the resultant couple must be counteracted by a couple produced by the reaction at the other bearing which can thus be determined.

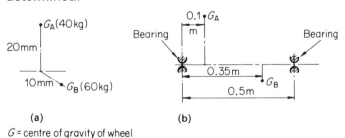

Fig. 9.4 Two views of the shaft and the wheel mass centres

G = centre of gravity of wheel

The mass-moment diagram is then drawn to include the couple produced by the reaction just determined. The resultant mass moment must be counteracted by that produced by the reaction at the first bearing. It is immaterial which bearing is chosen for the reference plane. When calculating couples, ω^2 is usually omitted as it appears in each term. Using the tabular method, and the plane of the left-hand bearing as a reference, the mass moments and couples are shown in Table 9.2.

Table 9.2.

mass (kg)	radius (m)	mass moment (kg m)	distance from LH bearing (m)	couple (kg m²)
40	0.02	0.8	0.1	0.08
60	0.01	0.6	0.35	0.21

From Fig. 9.5(a), *ob* is the resultant couple and scales 0.183 kg m². This must be converted into a true couple by multiplying by ω^2.

$$\therefore \text{True couple} = 0.183 \times \left(1200 \times \frac{2\pi}{60}\right)^2 = 2890 \text{ Nm}.$$

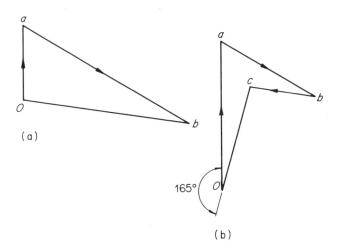

Fig. 9.5 (a) Couple diagram
(b) Mass moment diagram

Since the right-hand bearing is 0.5 m from the left, the dynamic reaction = $\dfrac{2890}{0.5}$ = 5780 N.

Its direction will be from *b* to *o* which is 82.5° anticlockwise from *oa* see Fig. 9.5(a). It will always remain in this relationship to *oa*. The mass moment diagram is now drawn (Fig. 9.5(b)) to include the mass moment derived from couple *bo* in Fig. 9.5(a), i.e. 0.183/0.5 = 0.366 kg m. From Fig. 9.5(b) *oc* is the resultant mass moment and scales 0.57 kg m. Multiplying by ω^2 converts this to a centrifugal force which produces the dynamic reaction at the left-hand bearing, i.e. 9001 N. The direction is $c \rightarrow o$ which is 165° anticlockwise from *oa*.

This is sufficient to answer the questions posed in Example 9.2 but it may be useful to examine the information further. The dynamic reactions of the bearings will vary in direction according to the angular position of the wheels and shaft which will give rise to vibration. In addition to these dynamic reactions the bearings will have static reactions due to the weight of the shaft and wheels. These are calculated by taking moments about one bearing. Hence, neglecting the weight of the shaft:

$$R_L = \frac{(40 \times 9.81 \times 0.4) + (60 \times 9.81 \times 0.15)}{0.5} = 490.5 \text{ N};$$

$$R_R = (40 \times 9.81) + (60 \times 9.81) - 490.5 = 490.5 \text{ N}.$$

To obtain the maximum bearing reaction when the shaft is in motion these static reactions must be added to the dynamic ones.

R_LMAX $= 9001 + 490.5 = 9491.5$ N, say 9.49 kN, occurring when *co* in Fig. 9.5(b) is vertically upwards.

R_RMAX $= 5780 + 490.5 = 6270.5$, say 6.27 kN, when *bo* in Fig. 9.5(a) is vertically upwards.

Remember that in this analysis we have been considering bearing reactions. The forces applied to the bearings, both static and dynamic, are numerically the same but opposite in direction.

9.4 Balancing non-coplanar mass systems

To achieve balance statically and dynamically, both the unbalanced forces and the unbalanced couples must be neutralised. The method generally adopted is to use just two balancing masses:

(i) one to balance the unbalanced centrifugal couples;
(ii) one to balance the unbalanced centrifugal forces.

As these balancing masses or their radii will initially both be unknown the effect of one is eliminated, temporarily. This is achieved by dealing with out of balance couples first. As the couples can be calculated from any suitable reference, the plane of one of the unknown balancing masses is chosen. This eliminates this mass from the couple calculation.

By drawing the couple polygon the resultant out of balance couple is found. Its equilibrant is provided by a suitable positioning of the other balancing mass. The couple effect is now eliminated and the system can be considered as coplanar in the reference plane and balanced in the way explained in Section 9.2. The mass balancing the couples, which will now be known, must of course be included. The following examples illustrate the method.

Example 9.3
A shaft suspended between bearings 0.5 m apart carries the equivalent of two masses of 8 kg and 12 kg at radii of 0.2 m and

0.15 m respectively. The planes of rotation of the masses are respectively 0.1 m and 0.4 m from the left-hand bearing. The masses have an angular displacement of 60° from each other. The system is to be balanced by two masses placed at 0.2 m radius in planes A and B, 0.2 m and 0.45 m from the left-hand bearing. Determine:

(a) the magnitude of the balancing masses;
(b) the angular position of the balancing masses relative to the 8 kg mass.

Two views of the system are shown in Fig. 9.6(a) and (b). A table of mass moments and couples is drawn up (Table 9.3), with plane A chosen as the reference plane. Plane B could also have been used. If the distances to the right of a reference plane are considered to be positive, then those to the left must be negative. The couple due to the 8 kg mass is thus negative. A negative couple is drawn in the opposite direction to a positive one. See Fig. 9.7(a).

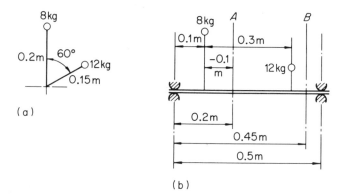

Fig. 9.6 Two views of the system

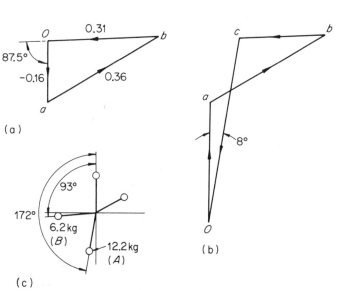

Fig. 9.7 (a) Couple diagram
(b) Mass moment diagram
(c) Positions of balancing masses

Terms in brackets in the table are derived from the couple diagram. From the couple diagram, Fig. 9.7(a), the equilibrant *bo* scales 0.3 kg m².

$$\therefore \text{Mass moment} = \frac{0.31}{0.25} = 1.24 \text{ kg m}.$$

This is inserted in the table.

$$\text{Mass } B = \frac{\text{mass moment}}{\text{mass radius}} = \frac{1.24}{0.2} = 6.2 \text{ kg}.$$

From the mass moment diagram, Fig. 9.7(b), the equilibrant *co* scales 2.44 kg m.

$$\therefore \text{Mass } A = \frac{2.44}{0.2} = 12.2 \text{ kg}.$$

The angular positions of the balancing masses are shown in Fig. 9.7(c)

Table 9.3.

mass (kg)	radius (m)	mass moment (kg m)	distance from reference plane (m)	couple (kg m²)
8	0.2	1.6	−0.1	−0.16
12	0.15	1.8	0.2	0.36
M_B	0.2	(1.24)	0.25	(0.31)

Example 9.4
A, B, C and D are four masses carried by a rotating shaft at respective radii of 40, 50, 80 and 60 mm. Their planes of rotation are 0.2 m apart. Mass B is 20 kg, C is 10 kg and D is 8 kg. For the complete balance of the system determine:
 (a) the mass A.
 (b) the angular positions of the masses.

Table 9.4 shows the couples and mass moments. Plane A is chosen as the reference plane as mass A (M_A) is unknown. The centre of gravity of mass B is assumed vertical, see Fig. 9.8.

Table 9.4.

mass (kg)	radius (m)	mass moment (kg m)	distance from reference. (m)	couple (kg m²)
M_A	0.04	(0.6)	Ref.	0
20	0.05	1.0	0.2	0.2
10	0.08	0.8	0.4	0.32
8	0.06	0.48	0.6	0.29

196 Mechanical Science IV

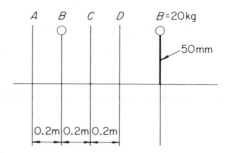

Fig. 9.8 Diagram of the masses and the shaft

The couple diagram, Fig. 9.9(a), is constructed, drawing *ob* vertical to represent the couple due to mass *B*. Then *bc* and *oc* represent the couples due to masses *C* and *D* which, for balance, must intersect at *c*.

Angle $\theta = 117°$ clockwise from *ob*.
Angle $\alpha = 258°$ clockwise from *ob*.

These give the angular positions of the masses *B*, *C* and *D* necessary for the couples to balance.

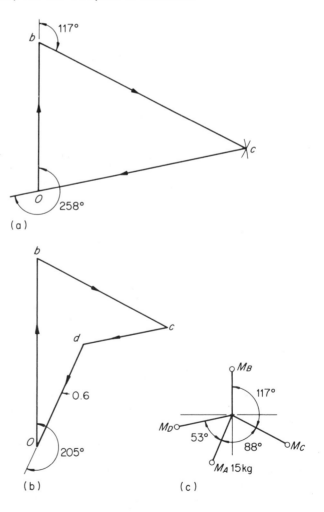

Fig. 9.9 (a) Couple diagram
(b) Mass moment diagram
(c) Angular positions of masses

With this information the mass moment diagram can now be drawn, Fig. 9.9(b). From the diagram, *do* scales 0.6 kg m and represents the mass moment due to the unknown mass *A*. As this is to be at a radius of 0.04 m,

$$\text{mass } A = \frac{0.6}{0.04} = 150 \text{ kg}.$$

Its angular position is 205° clockwise from *B*. The final positions are shown in Fig. 9.9(c).

9.5 Summary

(i) *Centripetal and centrifugal force* = $m\omega^2 r$
(ii) *Centrifugal couple* = $m\omega^2 r l$
(iii) *Static and coplanar balance*:

vector sum of mass moments = 0 or $\Sigma \overline{mr} = 0$

(iv) *Non-coplanar dynamic balance*:

vector sum of mass moments = 0 or $\Sigma \overline{mr} = 0$
vector sum of couples = 0 or $\Sigma \overline{mrl} = 0$

Exercises 9

Questions 1–4 concern theory and should be attempted initially without further reference to the text.

1 When a wheel supported in bearings is rotating at a constant angular velocity:
  (a) the linear velocity of any point on the rim is constant.
  (b) centripetal acceleration should be provided by the bearings.
  (c) centripetal force should be provided by the strength of the wheel material.
  (d) centrifugal force should be provided by the bearings.
  (e) the bearings should only be subject to the weight of the wheel.
  Indicate which of the above statements are correct.

2 A rotating system is completely balanced when:
  (a) it will come to rest in any position.
  (b) it will come to rest always in the same position.
  (c) in motion there is a constant bearing reaction.
  (d) in motion there is no resultant centrifugal force.
  (e) the centre of gravity of the rotating mass coincides with the axis of rotation.
  Indicate which of the above statements are correct.

continued

3 A non coplanar system of rotating masses is one:
   (a) where the masses have different planes of rotation.
   (b) where the bearings must exert a couple.
   (c) which is completely balanced by counteracting the couple.
   (d) which requires only two additional masses for complete balance.
   (e) which is balanced if each rotating mass is balanced.
   Indicate which statements are correct.
4 For a coplanar or a non coplanar rotating system the bearings:
   (a) must be capable of resisting the gravity force due to the mass of the system.
   (b) must always span the rotating masses.
   (c) will counteract any unbalanced couple.
   (d) will counteract any unbalanced centrifugal force.
   (e) are used as reference planes in the balancing calculation.
   Indicate which statements are correct.
5 Two coplanar masses of 15 kg and 8 kg rotate at radii of 0.2 and 0.15 m respectively attached to the same shaft. The plane of rotation is 0.2 m from one of the two bearings which are 0.3 m apart. When the 15 kg mass is vertically above the axis of rotation the 8 kg mass is 120° clockwise. Determine:
   (i) the out of balance force at 300 rev/min.
   (ii) for a speed of 300 rev/min, the maximum forces on the bearings and when they occur.
   (iii) the mass to be placed at 0.25 m radius and its angular position to completely balance the system.
6 A shaft, suspended in bearings 0.3 m apart, carries two gearwheels A and B. Gear A has a mass of 8 kg and B has a mass of 5 kg. Gear A is 0.05 m from the left-hand bearing and B is 0.2 m. The centres of gravity of the two wheels have eccentricities of 5 mm and 8 mm respectively. When the centre of gravity of A is vertically above the shaft axis, that of B is 240° clockwise. For a speed of 900 rev/min determine the maximum bearing reactions and when they occur.
7 A shaft, 1.15 m long, carries four eccentric masses A, B, C and D which are 0, 0.3, 0.75 and 1 m from one end. The respective masses are 8, 10, 15 and 6 kg and their eccentricities are 1.5, 2.5, 2.0 and 4 cm. Relative to the 8 kg mass, the centres of gravity of the other masses are respectively 50°, 210° and 270°. The bearings E and F have a span of 1 m, E being midway between masses A and B.

For a speed of 600 rev/min determine the maximum vertical forces acting on each bearing and when they occur.

8 A rotating shaft carries three discs A, B and C of mass 15, 25 and 20 kg respectively. The discs are spaced at intervals of 0.12 and 0.15 m from each other along the shaft. The eccentricities are 10, 12 and 18 mm respectively. To completely balance the system a mass, D, of 18 kg is placed in a plane 0.09 m from the plane of disc C, on the opposite side from A and B. Determine:
   (i) the angular positions of the mass centres of B and C relative to A;
   (ii) the radius of mass D and its angular position.

9 A shaft carries masses of 0.1, 0.1 and 0.045 kg in planes A, C and E respectively. The distances of the centres of gravity of the masses from the shaft axis are 38, 50 and 40 mm respectively and the angular positions of the masses in planes C and E relative to that in plane A are 42° and 327° clockwise. Determine the angular positions and magnitudes of the masses to be placed at radii of 50 and 38 mm in planes B and D respectively in order to balance the shaft completely.
   Reading from left to right, the distances between the planes are:
   ED = 100 mm;   DA = 125 mm;   AB = 75 mm;   BC = 150 mm.

10 Two rotors A and B are fixed to a shaft 0.3 m apart. The out of balance mass of A is equivalent to 0.1 kg at 100 mm radius and that the out of balance mass of B is 0.05 kg at 80 mm radius. The mass centres are 45° apart. Two balancing masses, C and D, are to be placed in planes C and D at 100 mm radius. The planes C and D are equally spaced between A and B. Determine the magnitude and angular positions of the masses C and D.

11 Four masses A, B, C and D are carried by a rotating shaft at radii of 40, 50, 80 and 60 mm respectively. Their planes of rotation are 0.2 m apart. Mass B is 2 kg, C is 1 kg and D 0.8 kg. Determine the mass of A and the angular settings of the masses to completely balance the shaft.

12 A rotating shaft carries four masses A, B, C and D. Mass B is 2.25 kg, C is 2.75 g and D is 2.18 kg. Mass A is unknown. The radii of rotation, in the same order are 50, 28, 25 and 40 mm. Mass C is 90° clockwise and mass D is 235° clockwise from mass B. The distances between the planes of rotation are A to B x m; B to C 0.25 m; C to D y m.
   If the shaft is in perfect dynamic balance determine:
   (i) the mass A and its angular position;
   (ii) the distances x and y.

Answers to Exercises

Chapter 1

1 Apart from showing the variation in shear force from section to section, Section 1.9 shows that the area of the shear force diagram between two sections gives the change of bending moment. For a uniformly distributed load the outline is a sloping straight line. Hence (a), (b), (c) and (d) are correct.
2 (a), (b) and (c) are correct. For verification of (a) see question 11 in the Exercises.
   (d) is only true for a beam with simple end supports.
   (e) is not correct, see Fig. 1.7.
3 (a), (b), (c) and (d) are correct.
4 (a), (c), (d) and (e) are correct. The maximum bending stress depends also on the section modulus,
5 (a), (b), (c) and (d) are correct (e) is incorrect as the support must also resist shearing.
6–14 inclusive. See Figs. 1.11 to 1.19

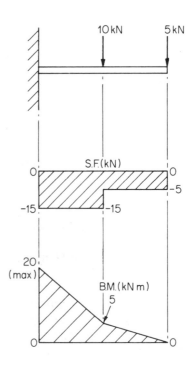

Fig. 1.11 Answer to question 6

Answers to Exercises 201

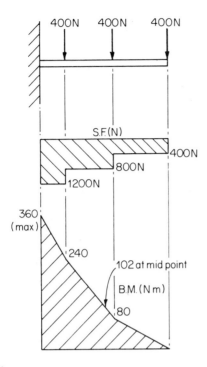

Fig. 1.12 Answer to question 7

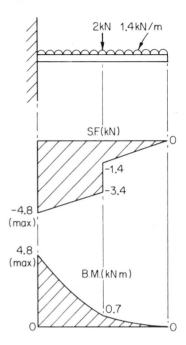

Fig. 1.13 Answer to question 8

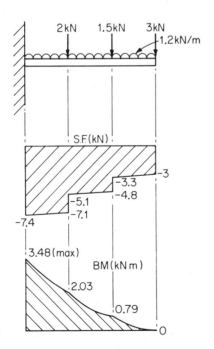

Fig. 1.14 Answer to question 9

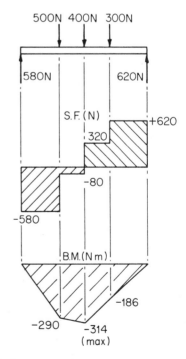

Fig. 1.15 Answer to question 10

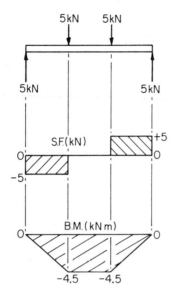

Fig. 1.16 Answer to question 11

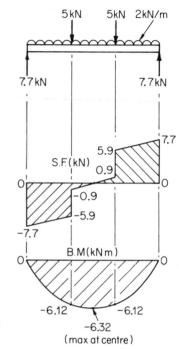

Fig. 1.17 Answer to question 12

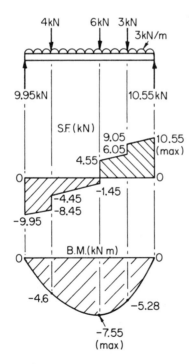

Fig. 1.18 Answer to question 13

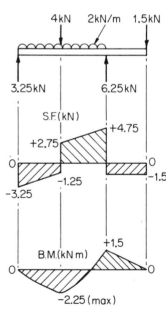

Fig. 1.19 Answer to question 14

15 (a) (i) $1.5x - 6.25[x-1] + [x-1]^2 + 4[x-2]$
(ii) $-3.25x + x^2 + 4[x-1] - 6.25[x-2] - [x-2]^2$
(b) 1.34 m from the RHE.
(c) 1.375 kN m.

16 (i) $w = 4x$ (ii) $F = 2x^2 - 0.375$ (iii) $\dfrac{2x^3}{3} - 0.375x$
(iv) 0.108 kNm, 0.433 m from the end of zero loading.

17 (i) 4.58 kN 5.42 kN (ii) 3.126 kN m, 1.3 m from the LHE.

18 Support reactions are 4.6 kN and 5.4 kN.
Concentrated loads of 4 kN 1 m from the LHE and 6 kN 2 m from the LHE. B. M. maximum = 5.34 kN m, 2 m from the LHE.

Chapter 2

1 (a), (b), (c) and (d) are correct; (e) is incorrect as symmetry about the neutral axis is not required. Examples are channel sections and T sections.

2 (a), (b), (d) and (e) are correct; (c) is incorrect as it is not true for channel sections and T-sections etc.

3 (a), (d) and (e) are correct; (b) is incorrect if the direct compressive stress is greater than the maximum tensile bending stress; (c) is incorrect since the eccentricity is measured from the centroid.

4 (a), (c) and (d) are correct; (b) is incorrect since the middle third area does not always appear as a rectangle; (c) is incorrect since the direct stress is considered uniform over the whole of the section.

5 (c), (d) and (e) are correct; (a) is incorrect since the maximum stress depends upon the position of the neutral axis relative to the outer layers; (b) is incorrect since only in special cases is the radius constant.

6 (i) 3.75 MN/m^2; (ii) 10 667 m.
7 (i) 82.5 mm; (ii) 4 880 cm^4.
8 (i) 27.5 cm; (ii) 150 938 cm^4.
9. (i) 1.294 m; (ii) 213 Nm.
10 (i) 81.5 MN/m^2; (ii) 30.7 m.
11 (i) 2.78 MN/m^2; (ii) 18 100 m.
12 3.115 m.
13 91.67 kN/m^2 tensile; 258.33 kN/m^2 compressive; zero axis 0.047 m from the centroidal axis.
14 (i) 77.8 kN/m^2 tensile; 246.2 kN/m^2 compressive.
(ii) 97 mm from the flange opposite the load.
15 (i) 364 mm; (ii) 442 mm.
16 (i) 300 kN/m^2 bending; 250 kN/m^2 direct.
(ii) 350 kN/m^2 tensile; 850 kN/m^2 compressive.

(iii) Line of zero stress cuts the shorter axis 125 mm from the centroid and the longer axis 167 mm from the centroid remote from the load point.

17 51.3 mm.

18 (i) 116.6 kN/m² tensile; 450 kN/m² compressive.
(ii) The line of zero stress cuts the shorter axis produced 111 mm from the centroid, and the longer axis produced 188 mm from the centroid, remote from the load point.

Chapter 3

1 (a), (b) and (d) are correct; (c) is incorrect since the slope refers to the loaded beam; (e) is not true for cantilevers.

2 (a), (d) and (e) are correct; (c) is only approximately correct; (b) is only constant for particular types of loading when the bending moment is constant.

3 (a), (b) and (d) are correct; (c) is incorrect as two integrations are necessary; (e) is incorrect as both upward and downward deflections occur.

4 (a), (b), (d) and (e) are correct; (c) is incorrect as the [ ] terms are conditionally used.

5 (a) and (e) are correct; (b) is incorrect as brackets are integrated complete; (c) is incorrect e.g. cantilevers; (d) is incorrect e.g. cantilevers.

6 (i) 0.215°; (ii) 2 mm; (iii) 0.625 mm.

7 1.56 m; 0.211°.

8 (i) 0.94 mm; (ii) 0.87 mm.

9 6670 cm⁴.

10 0.203 mm.

11 0.185°; 1.12 mm.

12 1330 cm⁴.

13 (i) 2.57°; (ii) 18.4 mm under 6 kN load; 12.2 mm under the 4 kN load.

14 (i) 0.46° at the LH support; (ii) 4.9 mm, 1.02 m from the RHE.

15 (i) $\frac{1}{33}(-x^2 + 0.533x^3 - 0.533[x-1.5]^3 - 2[x-1.5]^2 + 0.3)$

(ii) $\frac{1}{33}(-\frac{x^3}{3} + 0.134x^4 - 0.134[x-1.5]^4 - \frac{2}{3}[x-1.5]^3 + 0.3x)$

(iii) 3.82 mm (iv) 0.083°

16 (i) 1.62 mm (ii) 0.06° at the left hand support.

Chapter 4

1. (b), (c), (d) and (e) are correct; (a) is incorrect as reduction in cross-section will occur.
2. (a), (b), (d) and (e) are correct; (c) is incorrect as isotropic material must be homogeneous, but a homogeneous material need not be isotropic.
3. (a), (b), (d) and (e) are correct; (c) is incorrect as they can be both tensile or both compressive.
4. (a), (b), (c), (d) and (e) are all correct.
5. (a), (b), (d) and (e) are correct; (c) is incorrect since $E = 2G(1 + v)$.
6. (i) 55.3 mm; (ii) 50.5 mm.
7. (a) 53.5 mm; (b) 47.9 mm.
8. (i) 21.65 MN/m² tensile; 12.5 MN/m² shear; 25 MN/m² resultant
   (ii) 25 MN/m² tensile; 25 MN/m² compressive on planes at 45° to pure shear planes.
9. 37.4 MN/m².
10. (i) 13.7 MN/m² tensile, 37.6 MN/m² shear; (ii) 40 MN/m² 70°;
    (iii) 40 MN/m².
11. (i) 53.5 MN/m², 19.7 MN/m², 57 MN/m²; (ii) 20 MN/m².
12. (i) 33.7 MN/m² compressive, 6.7 MN/m² tensile;
    (ii) 24° for the tensile stress, 114° for the compressive;
    (iii) 20.2 MN/m².
13. (i) 24.7 MN/m² tensile, 44.7 MN/m² compressive; (ii) 111.2° and 21.2°; (iii) 37.2 MN/m².
14. (i) 1.44 MN/m² and 5.56 MN/m², both tensile; (ii) 38° and 128°; (iii) 2.06 MN/m².
15. (i) 3.75×10^{-6}; (ii) 7.5×10^{-5} cm³.
16. 3.375×10^3 mm³.
17. 80×10^9 N/m²; 33.3 GN/m².
18. 5.083×10^{-4} long.; -1.9×10^{-4} and -1.57×10^{-4} lat.; 19.9962×19.99686.
19. (i) 90 MN/m²; (ii) 45 MN/m²; (iii) 0.293 m³.
20. (i) 6.86 bar; (ii) 0.014 m³.

Chapter 5

1. (a), (b) and (c) are correct; (d) is incorrect as some slip can occur; (e) is incorrect compared with gears etc.
2. (a), (b), (c) and (e) are correct. In this case (d) is also correct in comparison with flat belts.
3. (a), (c), (d) and (e) are correct; (b) is incorrect and should refer to (velocity)².
4. (a), (b), (c), (d) and (e) are all correct.

5 (a), (b) and (d) are correct; (c) is correct since $T_{max} = 3T_c$ and (e) could be correct since m, the mass/unit length of the belt, depends upon the density of the belt material.
6 (i) 222 N; (ii) 348 N.
7 (i) 622 N and 311 N; (ii) 466.5 N; (iii) 4.98 kW.
8 (i) 108 mm; (ii) 417 N.
9 5.83 kW.
10 6.94 kW.
11 (i) 66.6 mm; (ii) 180°; (iii) 1.72 MN/m²
12 (i) 149°; (ii) 4.
13 (i) 174°; (ii) 236 N; (iii) 0.162.
14 (i) 0.473 m; (ii) $T_1 = 666.7$ N; $T_2 = 76.2$ N, 5 belts.
15 (i) 4502 rev/min; (ii) 21.24 kW; (iii) $T_1 = 500$ N, $T_2 = 49.4$ N.

Chapter 6

1 (a), (c), (d) and (e) are correct; (b) is incorrect since pairs can be pinned, sliding etc.
2 (a), (b) and (c) are correct; (d) is incorrect and (e) is incorrect, but the relative velocity will be at right angles to the line joining them.
3 (a), (b), (e) are correct; (c) and (d) are the wrong way round since for links rotating in the same direction the velocities are subtracted.
4 (a), (b), (c) and (d) are correct; (e) is incorrect since linear velocities are represented.
5 (a), (b), (c), (d) and (e) are all correct.
6 (i) *AB* 12.5 rad/s, *BC* 28 rad/s; (ii) (a) 0.506 m/s, (b) 0.194 m/s.
7 (i) 6.825 m/s; (ii) 0.2 m/s.
8 (i) *BC* 21 rad/s, *CD* 15.3 rad/s (ii) 7.6 m/s.
9 (i) 14 m/s (ii) *CD* 35 rad/s; *AD* 35.2 rad/s (iii) 0.59 m/s.
10 (i) *AB* 0.162 rad/s; *BC* 0.66 rad/s (ii) 5.57 m/s (iii) 1.0 m/s (join *a* and *c* on the velocity diagram.)
11 (i) 4 m/s; (ii) 13.67 rad/s.
12 (i) 2.3 m/s; (ii) 2.3 m/s; (iii) 0.355 m/s.
13 (i) 1.8 m/s; (ii) 0.59 m/s, 23°
14 (a) (i) 25 m/s; (ii) 6.5 m/s; (b) 1.87 m.
15 (i) 3.15 m/s; (ii) 2.7 m/s.

Chapter 7

1 (a), (b), (c) and (e) are correct; (d) is incorrect as the starting torque is initially increased.

2 (c), (d) and (e) are correct; (b) is correct for an engine or crank press; (a) is not strictly correct since time is usually converted to angular distance in radians.
3 (a), (b), (c) and (d) are correct; (e) is incorrect as the denominator should be *mean speed*.
4 3.35 Mg/m^3; 4.935 kJ.
5 14.18 rev/min.
6 (i) 14.8 kW; (ii) 304 kg.
7 1.92 kg m^2
8 (i) 3.32 kg m^2; (ii) 0.93.
9 (i) 6400 N m; (ii) 824.4 kg m^2.
10 (i) 6000 N m; (ii) 1360 kg; (iii) 0.75.

Chapter 8

1 (b), (c) and (d) are correct; (a) is incorrect as there are six; (e) is incorrect as there is rotation and translation.
2 (a), (b), (c), (d) and (e) are all correct.
3 (a), (b), (c), (d) and (e) are all correct.
4 (a), (b), (c), (d) and (e) are all correct.
5 (a), (b), (c), (d) and (e) are all correct.
6 0.2 m.
7 (i) 1.12 Hz; (ii) 1.23 rad/s.
8 (i) 0.293 Hz; (ii) 2.64 rad/s and 4.85 rad/s.
9 (i) Longitudinal 2.65×10^8 N/m, lateral 0.28×10^6 N/m, torsional 10.9×10^3 N m/rad;
(ii) 669 Hz, 21.7 Hz, 35.7 Hz.
10 (a) 0.414 kg m^2; (b) (i) 0.0495 kg m^2, (ii) 49.7 mm.
11 9.1 Hz.
12 (i) 4.418 mm; (ii) 100.4 kN.
13 30.8 to 32.08 Hz
14 126 Hz.
15 (i) 30 Hz; (ii) 31.17 Hz.
16 25.9 Hz.
17 (i) 1505 rev/min; (ii) 1492 rev/min.
18 1760 rev/min.
19 4304 rev/min.

Chapter 9

1 (c) and (e) are correct. The linear velocity of a point on the rim is changing constantly in direction, the bearing should only provide gravity forces.
2 (c), (d) and (e) are correct; (a) applies only to static balance but would be true for a coplanar system.
3 (a), (d) and (e).

4 (a) is correct; (c) and (d) are partly correct although the bearings are not designed to do so.
5 (a) 2566 N 23.6° clockwise from the 15 kg mass;
   (b) 1861 N RH; 930.2 N LH when 15 kg mass is 156° clock from vertical;
   (c) 10.4 kg 204° clock from 15 kg mass.
6 RH 259 N, 74° from given position; LH 340 N 158° from given position.
7 E 967 N 159° from given position; F 1074 N 216° from given position clockwise.
8 (i) B is 155°C 289° clockwise from A;
   (ii) D is 91° clockwise from A at 12 mm radius.
9 m_D is 0.105 kg 108°; m_B is 0.20 kg 218° both clockwise from A.
10 m_D 0.071 kg 308°, m_C is 0.176 kg 171° both clockwise from A.
11 m_A 1.5 kg 205° clockwise from B; C is 117° and D 260° clockwise from B.
12 m_A is 2.18 kg 201° clockwise from B; $x = 0.23$ m, $y = 0.33$ m.

(Hint—draw mass moment diagram first)

Index

angle of lap, 110
anisotropic, 71

balancing
  coplanar, 189
  couple, 190
  mass moment, 189
  non coplanar, 190, 193
  space diagram, 189
belt drives
  angle of lap, 110
  crossed, 110
  open, 110
  power transmission, 108
  power transmission maximum, 113
  pulley multigrooved, 106
  pulley v, 106, 115
  round belt, 116
  v-belt, 115
bending formula, 26
bending moment diagram, 1
bending moment
  relationship BM, SF and loading, 17
  standard cases, 6—8
bending theory, 1
  sign convention, 2
bulk modulus, 78, 98

centrifugal, centripetal force, 188
centrifugal of centripetal tension, 106
circumferential stress, 76
chain, 124
coefficient of
  energy fluctuation, 154
  friction (virtual), 116
  speed fluctuation, 153
complimentary shear stress, 79
contraflexure, point of, 17

coplanar
  balancing, 189
  stress system, 85

deflection of beams, 47
  Macaulay's method, 57
  standard cases, 53—57, 63, 66
Dunkerley's Formula, 178

eccentric loading
  compressive, 31
  double eccentricity, 37
  resultant stress distribution, 33
  tensile, 33
elastic constants, relationship between, 97

flywheels
  energy of, 146
  energy fluctuation, 151
  mean torque, 149
  radius of gyration, 146
  solid disc & annular, 159
  speed variation, 149
  turning moment diagram, 148

homogeneous material, 71
hoop stress, 76

inner dead centre, 136
isotropic material, 71

link, 124
  angular velocity of, 128
  relative motion of, 125
  slotted, 138

Macaulay's method, 57
mechanisms
   four bar chain, 130
   inversions, 125
   link, 124
   pair, 124
   pins, rubbing velocity, 129
middle third & quarter rules, 41
Mohr's stress circle, 88
modulus of
   elasticity, 98
   rigidity, 97, 98

neutral axis, 27
neutral plane, 27
non isotropic material, 71

orthotropic material, 71
outer dead centre, 136

pair, 122
parallel axis theorem, 27
pins, rubbing velocity, 129
poisson's ratio, 72
power transmission, 108
principal stress, 87
pulleys
   multigrooved, 106
   v, 106, 115

quadric cycle chain, 130

radius of curvature, 47
radius of gyration, 35, 146
relative velocity, 123

section modulus, 31
simple harmonic motion, 165
simple vibration, 165
single slider crank chain, 134
   inversions of, 137
short columns and struts, 41
slope of beam, 47
strain
   longitudinal, 72
   lateral, 72
   volumetric, 74
stress
   compressive, 70
   shear, 70
   tensile, 70
   complex compressive, 82
   complex shear, 83
   complex tensile, 82
   complimentary shear, 79
   maximum shear, 87
   Mohr's circle, 88
   principal, 87
stresses in thin cylinders and spheres, 75
   hoop, 76
   longitudinal, 76

thin cylinders and spheres, 75, 76
   change in volume, 75
torsion formula, 93
turning moment diagram, 148
   for engine, 148
   for a press, 149
   speed variation, 149
   mean torque, 149

vibrations
   beam with single concentrated load, 173
   beam with distributed load, 175
   damping of, 164
   degrees of freedom, 163, 164
   Dunkerley's formula, 178
   lateral, 171
   longitudinal, 168
   simple, 164
   torsional, 169
virtual coefficient of friction, 116
volume change, thin cylinder or sphere, 77
volumetric strain, 74

whirling of shafts, 181
   shaft with single mass, 182
   speed, critical, 182, 184